T0348840

Breakthrough Food Product Innovation
Through Emotions Research

Breakthrough Food Product Innovation

Through Emotions Research

David Lundahl

AMSTERDAM ● BOSTON ● HEIDELBERG ● LONDON
NEW YORK ● OXFORD ● PARIS ● SAN DIEGO
SAN FRANCISCO ● SINGAPORE ● SYDNEY ● TOKYO

Academic Press is an imprint of Elsevier

Academic Press is an imprint of Elsevier
32 Jamestown Road, London NW1 7BY, UK
225 Wyman Street, Waltham, MA 02451, USA
525 B Street, Suite 1800, San Diego, CA 92101-4495, USA

Notice
No responsibility is assumed by the publisher for any injury and/or damage to
persons or property as a matter of products liability, negligence or otherwise, or from
any use or operation of any methods, products, instructions or ideas contained in the
material herein. Because of rapid advances in the medical sciences, in particular,
independent verification of diagnoses and drug dosages should be made

British Library Cataloguing-in-Publication Data
A catalogue record for this book is available from the British Library

Library of Congress Cataloging-in-Publication Data
A catalog record for this book is available from the Library of Congress

ISBN: 978-0-12-387712-3

For information on all Academic Press publications
visit our website at www.elsevierdirect.com

Typeset by TNQ Books and Journals

Printed and bound by CPI Group (UK) Ltd, Croydon, CR0 4YY

Transferred to digital print 2012

Working together to grow
libraries in developing countries

www.elsevier.com | www.bookaid.org | www.sabre.org

ELSEVIER   BOOK AID
            International   Sabre Foundation

Contents

Dr. David Lundahl is an expert in applying consumer and marketing research to product innovation and development, bridging the gap between brand and consumer through a behavior-driven approach to innovation. Trained as a statistician and a food scientist, he is known for his statistical applications to product development. Yet, his passion for innovation has led him down a unique path. He brings to innovation teams a diverse set of disciplines, including innovation process management, marketing research, and behavior psychology.

As a consultant and educator, he has applied his expertise to help companies bring to market successful innovative products within the consumer packaged goods (CPG) industry.

Lundahl was a founding faculty member at the Oregon State University Food Innovation Center in 1996. He chaired the Sensory and Consumer Sciences Division of the Institute of Food Technology in 2000. Further, he has taught numerous workshops on applying consumer research to product innovation and development, has written numerous publications, and has been a keynote speaker at various industry conferences.

In 2003, Lundahl founded InsightsNow, which was recognized in 2010 as one of the fastest growing US-based marketing research companies. His vision to change the game through innovation has led to the formation of a new behavioral framework for breakthrough food product innovation. He lives in Corvallis, Oregon, with his wife and three daughters.

Preface

Innovation has been a focal point of my professional career since the early 1990s. Having received a PhD in Food Science & Technology and a MS in Statistics from Oregon State University, I was well acquainted with science and the scientific method. However, I quickly learned that the application of scientific methods is not the same in industry as in the university. Successful innovation requires learning that often goes beyond science and scientific methods. I found many of the scientific methods useful in the university environment to be inadequate in the industry. They often lacked insights that inspired creativity and did not factor in human intuition to drive decisions.

My professional career started in 1989 as a researcher for Frito-Lay within their Research and Development division. At Frito-Lay, I worked within the consumer product research team. This team pioneered a number of methods in the 1980s to automate consumer product research. The efficiency and effectiveness of this program was so successful that it propelled Frito-Lay into an industry innovation leadership role. In 1991, I was assigned to the Baked Lays product development team where I experienced (first hand) the innovation case study discussed in Chapter 2. My experiences at Frito-Lay laid the foundation for many of my beliefs about the role of research in innovation and product development.

In the course of these past 20 years, I have seen technology impact the food industry in a number of ways. Certainly, the speed and diversity of research information has opened up new vistas for how to bring new and improved products to market. However, technology has also fundamentally impacted the way consumers behave and their speed of change. Consumers now have immediate access to information that shapes how they select, seek, share, and react emotionally to brands, products and product experiences.

The food industry is struggling to adapt to this consumer change. In the early 2000s, it became clear that researchers were falling behind in delivering insights of strategic value to decision makers. In 2005, the value of emotions research began to take shape as a key to deepening insights of more strategic value. My colleagues at InsightsNow and I began to explore more holistic research approaches to understanding the "whys" of consumer behavior. We began to piece together a new framework that could integrate a mix of qualitative and quantitative research information. This led to the early ideas behind this book – the emergence of emotions research.

In 2007, behavioral psychology began to take center stage as a key contributor to understanding consumer market behavior. Behavioral economics, with its roots in behavioral psychology, was being applied in different vertical markets to "nudge" people along in their behaviors. However, behavioral economics had yet to find applications in the food industry. This all changed with the recession that impacted the food and other consumer packaged goods industries in the fall of 2008. Change in consumer behavior occurred so rapidly that no one in the food industry was spared. It was a clear signal that emotions were a major factor in driving market change. It was clear that the time was right for a new, behavior-driven approach to food product innovation.

In mid 2009, I formed a team of industry colleagues to provide critical thinking around the ideas that inform this book. Over a 22 week period, they provided critical feedback on the materials presented within each of the 11 chapters for this book. This team included Dr. Mina McDaniel, Simon Chadwick, Bill Graves, Frank Hall, Greg Stucky, Norm Galvin, and Alec Maki. Mina was my advisor for my PhD program and is a world-wide expert in sensory science. Simon Chadwick is the former Global Chief Executive Officer for NOP and a leader in the marketing research industry. Bill Graves is the former Senior Director of Consumer Insights for Kraft Foods. Frank Hall is an expert in business development and a consultant in technology and innovation. Greg Stucky, Norm Galvin and Alec Maki are colleagues at InsightsNow. Greg is our Chief Research Officer, Norm our Executive Vice President, and Alec our VP of Product Development.

Once the material was gathered for writing this book, it took another year of "baking" to get it right. I have been very impressed with the professionalism and editorial contributions from those at Elsevier. It is my hope that you will find this book stimulating, and disruptive, and that it will set a path ahead for you to change the game in how food products are developed from ideas into breakthroughs.

Acknowledgments

I must acknowledge all the members of the industry team who contributed their time to provide critical thinking to the content in this book. Among this group, I highlight the contribution of Greg Stucky. He and I have worked together since 1996 – first at InfoSense and now at InsightsNow. His insights have been extremely helpful in thinking about how innovation teams learn – how insights and information must carry over from phase to phase. He has also contributed to thought on how to achieve insights for incorporating product cues that signal emotional impact. Further, his insights have contributed to thought about the importance of achieving harmony between the components of products.

I also acknowledge Dave Plaehn, who has worked with me since 1997, first as my Research Associate when I was still a professor at Oregon State University and then over the years we have worked together at InfoSense and InsightsNow. Dave is the chief contributor to a number of innovative analytic methods which I have had the pleasure to publish with him over that period. I also acknowledge the contribution of Doug Hansen – a graphic artist – and Alec Maki, who have been instrumental in translating complex ideas into a number of outstanding graphical images used throughout this book.

Finally, I must thank my family for putting up with my many long hours spent writing this book. This includes my wife, Shelly, and our three children, Erika, Johanna and Christina. At the time this book was written, Erika was a junior studying Writing at Ithaca College in New York. She was very helpful in contributing to the editing of the book chapters.

David Lundahl

Chapter 1

Change in the Food Industry

If you don't create change, change will create you.

Anonymous

A TIME OF CHANGE

Never before in the history of human civilization have we seen such dramatic change in the way that people are able to connect to each other and to consumer brands. The landscape of consumer product markets has been reshaped – enabled and fostered by information technology that has changed the way that consumers seek and share their experiences with their peers. Consumers are rapidly changing in how they come to know, purchase, and consume food products.

In a few short years, we have seen the product ownership balance of power shift from brand owner to consumer. Brand owners are no longer able to drive demand through mass media marketing tactics. They are waking up to a world where peer-to-peer sharing has amplified the desires of the individual. Consumers have discovered their new power – a power to rapidly self-mobilize and place new demands on brand owners. Now sitting on top of the "food chain" is an empowered consumer, seizing opportunity in a new, proactive role to drive market change.

Product marketers, developers, creative chefs, package designers, marketing researchers, sensory scientists, and business managers are all striving to adapt to this change in their respective roles. Companies are seeking better ways to organize and streamline their innovation to sustain a competitive edge in this new environment.

This book is about innovation in the face of this change. In the chapters ahead, a more behavior-driven approach to innovation will be presented. This new approach shifts the focus of innovation from the product to the experience. It reshapes innovation from a linear to an iterative process driven by insights, based upon the science of emotions to motivate consumer behavior – delivering outcomes that consumers seek.

Societal Change

Social networking and social media are having far reaching effects on individual behavior; and the newfound power of the individual is changing society.

Breakthrough Food Product Innovation Through Emotions Research. DOI: 10.1016/B978-0-12-387712-3.00001-3

Western culture, in particular, has been radically changed through social media with the formation of thousands of loosely held, virtual social communities held together by shared values, interests, passions, and beliefs.

These virtual communities are significantly altering consumer attitudes and values. Peer-to-peer sharing within virtual communities has increased awareness, enabling a much more informed consumer to take action. The millennial generation (also known as NetGens) are characterized as being more value-driven and their use of social media is having a dramatic influence on consumer markets. This generation includes 28% of Americans (88 million) born[1] between 1980 to 1995, compared to the 76 million Baby Boomers born between 1946 and 1960. We are just now seeing the impact of this generation on our culture. Eric Qualman, author of the bestseller *Socialnomics* (2009, John Wiley & Sons), notes that 97% of NetGens have joined a social network. One out of nine couples in the United States met through a social network. If Facebook were a country, it would be the world's fourth largest.

As more and more people join in the social networking revolution, it is changing where people get information, how fast they get information and what sources they trust. Peer-to-peer information is now much more trusted than information sourced from brand owners.[2] The impact of these changes in seeking and sharing behavior is also leading to a much more emotionally-driven consumer.

According to social psychologists, consumers are experiencing a sense of "depletion" in the current marketplace, having less time, energy, and financial resources. This creates more emotional and impulsive buying habits. Further, the speed by which information is shared on a global scale is resulting in dramatic shifts in consumer awareness, attitudes, and behaviors. The relationship between consumer, brand, and brand owner is more dynamic, ultimately impacting how consumers purchase products and change markets.

Consider what happened to Unilever when the Axe brand "women falling at the feet of men" commercial was aired in 2007 after their Dove brand had been running its "self-esteem, real beauty" campaign for three years. Consumer reaction was swift. Within 24 hours, the Internet was filled with YouTube™ videos and posts on blogs, social communities, consumer forums, and message boards venting anger at Unilever's hypocrisy – culminating in a special report by CNN.[3]

Market Change

These societal changes are having a dramatic effect on consumer product markets. The growing surge of highly engaged, demanding, and emotionally-driven consumers is contributing to massive market fragmentation and

1. http://en.wikipedia.org/wiki/Demographics_of_the_United_States#Population_growth

2. http://www.edelman.com/trust/2009/docs/Trust_Book_Final_2.pdf

3. http://www.youtube.com/watch?v=dRNbZQ7K3vo&feature=player_embedded

dynamics. In 1980, Alvin Toffler (*The Third Wave*, William Morrow, 1980) coined the term "prosumer" in predicting a future where the distinction between producer and consumer is blurred. Welcome to our world today! More than 20 years later, Dan Tapsott and Anthony William redefined "prosumer" in their book *Wikinomics* (Portfolio, 2006) to mean a consumer who is more proactive in co-creating goods with manufacturers. In this book, a more behavioral definition of "prosumer" is introduced. This definition includes not only the typical consumer behaviors of selecting and consuming foods, but also the behavior of "seeking out" brands that consumers believe will work for them and the "sharing" of their discoveries with peers.

Understanding this behavioral definition is essential to all who are involved in product innovation and development. The behavior of seeking out and/or sharing is leading to a more dynamic marketplace. The speed by which opinions can be shared and information sought leads to markets that are becoming more fragmented. This fragmentation is accompanied by market churn, where loyalties of the past are eroding. This also opens up opportunities for companies to built new brand loyalties through a wide range of niche markets.

As populations grow and new social media-driven trends emerge, many of these fragmented, dynamic markets (i.e. niche markets) will no doubt form the basis for larger, main-stay markets in the future. Therefore, it is extremely important that corporations, as brand owners, begin planning for long term corporate sustainability. Consumer awareness of corporate behaviors will have long-lasting impact on how companies, the brands they own, and their messaging are trusted by consumers. Now, more than ever, products released under specific brands must perform against consumer expectations – to deliver on the promises of the brand.

This new world is now, more than ever, a brand-driven marketplace, driven by social issues such as social justice, sustainability, health and wellness, and a wide range of other issues including safety, longevity, and comfort. How consumers come to know brands, trust what they say and align with these issues will likely impact the success of brands and brand owners.

Commoditization

At the same time as food companies are trying to adapt to huge changes in society and markets, they are also facing steep declines in margins due to commoditization. Consumers are besieged with a myriad of product choices in retail stores. That consumer product selection is influenced by a breadth of factors ranging from thinning wallets, to an increase in the number of alternatives on the shelves, to how those choices will affect their lives. Consumers want products that will make their lives better at the lowest possible price, whether it's food, beverages, personal care products, or products for their homes.

CEOs, marketers, and researchers are finding it increasingly difficult to become differentiated in the eyes of the consumer. Consider the case of a brand owner marketing products within the canned beans category. How do you differentiate yourself from your competition? What *does* differentiate one can of beans from the next in the mind of the consumer? Traditionally, differentiation has come through assessing the needs of consumers and applying innovation to deliver features, functionality and sensory qualities that address these needs. However, this approach to innovation has not proven sufficient to break away from the chains of commoditization that grip most brands and squeeze margins. With pinched pocketbooks and a flood of competitive, quality store brands, we are seeing increased erosion in the bulwark that main brands have traditionally built.

With commoditization, one brand has no features that differentiate it from the other brands and as a result, the consumer makes decisions based on price alone. It is one of the most difficult marketing challenges a marketer faces. With commoditization comes increased competition, and the lack of differentiation creates a downward cycle of lowered consumer desire to seek new functionality or "new and improved" intrinsic qualities. Economic downturn and market complexity leads to consumers entering the marketplace with reluctance to pay for what they see as unnecessary features and qualities. Brand owners end up with extreme price sensitivity and the nightmare of increased pressure on margins. As a result, brand owners are reluctant to invest in the creation of new and improved products, settling for cost-cutting as a way to offset lower margins. In the end, both brand owner and consumer suffer: brand owners realize lower margins, consumers less value.

This pervasive trend has hit the food industry particularly hard. The rise of store brands and private labels by retailers has further exasperated this situation for both the consumer and brand owner. With this trend, retailers are now effectively competing with brand owners - their own suppliers. No other industry sees this pervasive a problem in product development. Consider some of these statistics:

- 25% of all food and beverage purchases are store brands (NPD Group Research Reports, 2009).
- 97% of all households purchase store brands (NPD Group Research Reports, 2009).
- 66% of respondents have purchased store brands in the past month (Consumer Reports, May 2009).
- Store brands are priced competitively, averaging 30% lower in cost (IRI, 2009).
- Store brand product quality is believed to be equal to name brands (Meyers Research, 2005).

These statistics paint a picture of an industry in turmoil. The food industry is struggling to find a new formula for operation, a new business model for innovation that breaks the downward spiral of commoditization, creating a new cycle of profitability for brand owners and value for consumers. Economic downturn in general, combined with a lack of product success, has created an uncomfortable position for the food industry.

ADAPTING TO CHANGE

Storm Clouds on the Horizon

In 1992, the food retail industry appointed the Efficient Consumer Response (ECR) Commission[4] to assess the economic impact of product failure. They reported that the cost development and introduction of failed new food products was nearly $14 billion in the United States alone, twice as much as the profits from the 15 largest food companies that same year. Out of their analysis, the ECR identified four key areas as essential for getting the food industry back on track, and an efficient process for new product development was cited as most important.

However, in spite of numerous attempts to streamline and improve the innovation and development process, the spiral of increased failure continues. New product introductions of food and non-food items peaked in 2005, with a trending down to 2004 levels in 2008.[5] This downward trend reflects the industry's reluctance to bring new products to the market and risk failing.

Another dramatic shift over the past 10 years is the proportion of new food product introductions that reach sales of over $20 million in their first year has dropped from 13% to 7%. In non-food categories, a drop from 22% to a low of 5% of all introductions has been observed. It is not surprising that as margins erode and success rates drop, we are seeing fewer new products introduced.

During the past three years, food industry pacesetters – companies having products with greater than $20 million in first-year sales with nationwide distribution – are experiencing greater return on investment with line extensions versus new product introductions. This may be an indication of an increased focus on niche markets instead of introducing new products into broad-based markets. Irrespective of what is at the root of these statistics, it cannot be denied that the food industry has been struggling for some time to create sufficient value for consumers to break the chains of commoditization, to discover new ways to reconnect with consumers and to create a new cycle of value growth.

The Perfect Storm

These collective changes form the basis for a "perfect storm" for industry change. Swept up in this storm are many corporate ships flying the flags of thousands of brands. How these corporations navigate the stormy waters of

---

4. http://www.fmi.org/media/bg/?fuseaction=ecr1

5. Information Resources, Inc., 15 Years of New Product Pacesetters: Excellence in Innovation Drives CPG to the Next Level, 2008

societal and market change will depend on how they innovate and develop – i.e. how they change their approach to serving the demands of these highly proactive, engaged consumers.

These stormy times have led to a new age of corporate awareness. The food industry is finding that traditional approaches to innovation and development are too slow and inflexible to respond to these marketplace dynamics. Companies are beginning to understand the need to become more socially driven, to see the building of corporate trust as a critical goal. Corporations are beginning to view their brand trust as a key asset and to apply this brand trust to transform from a "goods manufacturer" strategy to a "brand manufacturer" strategy. More niche products are being made to serve a wide range of dynamic, fragmented markets. More companies are discovering the importance of emotional branding and focusing innovation on sustaining relationships with their consumers. More emphasis is starting to be placed on design and development to deliver product experiences with emotional impact.

However, there are still fundamental gaps between the brand owner and consumer. The strategies that companies are taking lack a fundamental understanding for what is truly driving consumer behavior. Few frameworks exist that help brand owners, marketers, marketing researchers, designers, developers and sensory scientists to understand how to innovate with the important end outcome of innovation, i.e. consumer behavior, in mind.

Although awareness is the first step in making significant strides to adapt to change, most companies still appear caught with one foot in the past, where

FIGURE 1.1   The gap between consumer and brand. (Please refer to color plate section)

innovation and development were conducted using linear, inflexible organizational processes, and where products were developed in bits and pieces within silos of domain experts. Companies are finding that these business practices no longer deliver a consistent flow of product successes. They are simply too slow, expensive, and inflexible.

To meet the short term, bottom line needs of food industry shareholders, companies are continuously reorganizing, seeking ways to cut expenses to maintain acceptable profitability. In reality, the resulting churn in employees is dramatically decreasing the knowledge capital retained by food companies. As historical knowledge gained through experience dissipates, the gap widens between brand owner and prosumer.

However, in these darkest of times, there is a silver lining ahead. *Wikinomics* described this storm as "mass collaboration," characterizing the far reaching effects of technology change on not only consumers, but also on a wide range of emerging business processes and practices. The food industry, historically slow to embrace change, is beginning to show signs of applying mass collaboration to innovation. Innovation leaders in the food industry are starting to speak openly about the need for open innovation to increase the number of new ideas that have commercial viability.[6] Mass collaboration is now leading the way forward through a wide range of new technology-enabled tools and services that facilitate more easy-to-use, inexpensive and effective forms of engagement with consumers. In the chapters ahead, a number of case studies will show proof that the food industry is awakening to how it can apply these tools and services to design and develop new foods and packaging.

However, technology alone cannot provide the new way forward out of the stormy waters of societal and market change. What is needed is a new paradigm for innovation that is enabled not just by technology, but by a new framework for innovation and development that is rooted in the science of consumer behavior. Only by combining technology with a basis of understanding of consumer behavior can the consumer packaged goods (CPG) industry truly adapt to change. This new paradigm for innovation and development must be aligned with the way consumers are seeking to change their world through their proactive consumer behaviors and also prove flexible enough to adapt to the dynamic marketplace.

The surviving corporate ships out there are the ones who are able to embrace this change as an opportunity to build and sustain new relationships with consumers. By embracing the reality of the seeking and sharing consumer, new innovation leaders are emerging who seek to engage their loyal consumers to help drive innovation. They are moving from innovation models focused on building products with different features and functionality to creating consumer experiences that consumers seek and are motivated to share. They are reaching

---

6. http://www.foodnavigator.com/Science-Nutrition/Open-innovation-Food-industry-needs-
    better-strategies-says-review

consumers through a new marketing mix that involves less mass marketing and more consumer collaboration, through digital marketing involving building of online communities. They are beginning to discover how mass collaboration changes everything – consumer, societal norms, markets and approaches to innovation.

Breaking Into the Clear Blue

Commoditization is only a symptom of a bigger problem: the food industry has not been able to find a formula to consistently bring to market products that break through the clutter. Escaping from these stormy waters into the clear blue requires a transformation in how to innovate in the face of change. Name brand and store brand owners (i.e. retailers) both need innovation – not in a competitive sense, but in a market sense. Innovative products are needed to re-energize stagnant product categories. Consumers have quickly learned that store brands are better – equal in value and lower in cost. However, it is reasonable to expect that the pendulum of interest in new and novel products will return. Consumers are by nature restless for newness; they will again be seeking new experiences and to build new relationships with brands that will fulfill their desires.

The emotional nature of consumers requires a different approach to innovation that makes operational the science of consumer behavior for food product innovation. This approach must embrace research methods that tap into consumer tendencies to seek relationships with brands through products and to share those experiences with peers. Such an approach is provided in this book. The research methods are more emotions-based. They employ techniques that are integrated and holistic – using a mix of qualitative and quantitative techniques to gain insight into how the various pieces and parts of products interact to form consumer experiences. This approach leads to insights through research that inspires and guides innovation teams – fostering their collaborative intelligence that uses their multidisciplinary wealth of knowledge.

This approach leads to a roadmap for innovation success in the food industry. It defines a new arena where brands and products are differentiated, using emotions as the beacon to help navigate the stormy waves of today's marketplace. Lighting the way will be emotions insight gained through research methods, based upon the science of emotions – uncovering the "whys" of behavior, inspiring innovators to imagine new consumer experiences and guiding managers in their innovation decisions.

Many stories are emerging for how this approach is leading companies into the clear blue. Some are described in this book as case studies gleaned through the public domain and some from published and unpublished research conducted by InsightsNow. Chapter 2 is about innovation and how it has changed over the past 25 years. Chapter 3 discusses the strengths of innovation teams and the challenges to them in applying their collective knowledge to overcome obstacles to innovation in the face of change. In Chapter 4, a summary of key knowledge about the science of emotions is reviewed and applied into a framework for emotions research. Chapter 5 then elaborates further on emotions research methodologies. Chapters 6 to 10 apply the framework to specific phases of the innovation and development process. Lastly, Chapter 11 completes the story of how the science of emotions can be applied to market launched food product innovations. The chapter concludes with a discussion of how the science of emotions leads to new possibilities to build, manage and retain knowledge about consumers through tracking the performance of launched food product innovations.

In many ways, this book is about transforming the research industry to enable more rapid learning-leading to more inspired creativity and accurate guidance in decision-making throughout the innovation and development process. The key to innovation, in the face of change, is speed in learning – achieved through a more flexible, dynamic approach to research enabled by real-time technology and a framework that integrates the resulting information into knowledge based upon the science of consumer behavior.

Key Points

- Due to social media, people are connecting in a more inter-related way than ever before.
- Consumers have more power in this social environment to influence markets and make decisions — companies MUST innovate in the face of this storm.
- Consumers are now more emotionally driven in how they behave in the marketplace, driven by depletion of resources and the emergence of prosumers — proactive consumers who are motivated to seek out brands that will work for them and share their discoveries with their peers.
- The emotionally-driven consumer has led to a fragmented, dynamic marketplace.
- The inability of the food industry to adapt to consumer change has led to commoditization — evidenced by the rise of store brands with cost as the only differentiator.
- Breaking through the clutter in the food industry requires a different approach to innovation that makes operational the science of consumer behavior for food product innovation.
- This book defines a roadmap for companies to compete in a new arena where brands and products are differentiated using emotions as the beacon to help navigate the stormy water of today's marketplace.

Chapter 2

Innovation

Innovation is the process of turning ideas into manufacturable and marketable form.

Watts Humprey

INNOVATION SUCCESS AND FAILURE

Frito-Lay Baked! Lay's

As early as 1991, Frito-Lay began working on developing a new, "better-for-you" potato chip. Frito-Lay's potato chips is a brand with almost universal recognition in the United States, and the leader in the snack food category. The crunch and flavor were a desired favorite for millions of consumers. Yet, potato chips were 30% oil or liquid fat by weight and processed by frying potato slices in large vats of oil. Frito-Lay marketers saw a concerning trend amongst consumers. Nine out of ten people were selecting "better-for-you" foods and were willing to pay more to get them.[1] Fatty foods like conventional potato chips were not perceived by consumers as "better-for-you".

To produce a potato chip without "bad-for-you" oil was a challenge. Essentially, Frito-Lay was tasked with taking a tasty potato chip that derives its flavor and mouth-feel from deep fat frying and making it healthier by minimizing the percentage of oil used. In other words: Make a potato chip that tastes like it was fried, but without frying it.

After more than four years of research, Frito-Lay offered early prototypes of a "baked chip" to test groups of consumers. The results in August of 1994 were unacceptable – it was dry and tasted like cardboard. It was clear that to give rise to a tasty and healthy new chip, Frito-Lay needed to further develop their process. By the end of 1995, consumer research reported a chip that was ready for the market in both texture and taste (Figure 2.1).

In 1996, the baked potato chip was introduced as Baked! Lay's®. For Frito-Lay, it was a chart breaker. The test market results returned the most positive response to a product introduction ever experienced by Frito-Lay. In the first

---

1. R. Mathews, Efficient New Product Introduction, July 1997, Supplement to Progressive Grocer, pp. 8–12

Breakthrough Food Product Innovation Through Emotions Research. DOI: 10.1016/B978-0-12-387712-3.00002-5

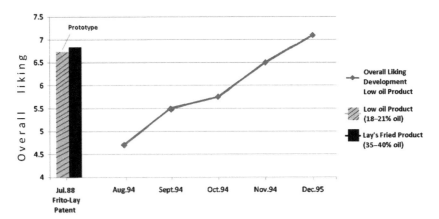

FIGURE 2.1 Scores of product Liking from early stage prototypes of Baked Lays® as adapted from US patent application (Adapted from Dreher, M. et al. 1988. Process for Producing Low Oil Potato Chips. Assignee US Patent 4,756,916. Frito-Lay, Inc., Dallas, TX) and from reported scores throughout the product development process prior to market lanuch (Adapted from Mathews, R. 1997. "Efficient New Product Introdution," July 1997 Supplement to Progressive Grocer, pp. 8–12).

year, the product grossed over $100 million in revenue and they experienced a 25% overall brand sales growth.[2] More than a decade later, the Baked! Lay's brand is still a $100+ million a year revenue product.

This case study is an example of innovation. While invention is the creation of something new, it does not ensure success. Inventions are seen in every area of industry: in product development as new products; in the technical research and development (R&D) centers as a new process or basis for a new product; and in universities or research centers as new ideas. Inventions may be scientific breakthroughs, but it takes innovation to make that something new into a market winner. *True innovation is the process of taking of an idea and making it a commercial success.*

In the case of Frito-Lay, marketers applied their domain expertise to understand how consumers were changing attitudinally and behaviorally. Researchers were tasked with the role of seeking out inventions that could be applied to solve the problem of developing an acceptable potato chip with little or no oil. At every step of the product design and development process, consumers were involved to help guide the innovation and development process. The innovation process started and ended with the consumer.

Orbitz: A Failure in Innovation

In 1994, Clearly Canadian Beverage Corporation had only $66 million in sales, compared to $141 million in 1992. In an effort to boost their slumping sales,

---

2. Ibid.

the company sought to apply innovation and develop new products. Their introduction of the Orbitz drink in 1996 will live in the minds of many consumers as one of the great product failures of all time. Touted as a "texturally enhanced alternative beverage," Orbitz contained tiny edible gelatinous gum balls floating in a fruit-flavored beverage. The industry initially hailed it for its inventiveness and it received numerous accolades for its design:

- 1996 *Mobius Award* in recognition of the outstanding package design.
- 1996 *Clear Choice Award* from the Glass Packaging Institute.
- 1996 *BevNet! Beverage Awards* for best packaging and appearance.
- Named the *"Hot drink of '96"* by *Rolling Stone Magazine*.

In spite of this great inventiveness, it was those very gelatinous gum balls that ultimately fast-tracked the product off the shelves, due to poor sales.[3] Within two years, the product was no longer commercially available. Instead of boosting sales, millions were lost in development and launch promotion. What went wrong?

Obviously, with any innovation there are risks. Innovation is the lifeblood of organizations; without innovation, companies experience missed opportunities. But there is no certainty that introducing a new or improved product to the market will bring success. It is important that companies recognize the reasons for both product failure and success. There are three generally accepted reasons:

1. How well the innovation is **founded** with regard to corporate strategy.
2. How well the innovation is **directed** by marketing research information.
3. How the innovation is **differentiated** in the mind of the consumer.

In the case of Baked! Lay's, the product line was well founded. The company had done its homework in understanding consumer desire to consume "better-for-you foods." The iterative product development process was guided by consumer product research into how well prototypes were liked. In addition, the resulting product launched into the marketplace was highly differentiated on the basis of being the first low fat, enjoyable potato chip product on the market.

In the case of Orbitz, Clearly Canadian had not accurately identified a purpose for Orbitz in the marketplace. While the product was highly differentiated with regard to product features (i.e. the floating balls in the liquid), consumers who tried the product did not want to repeat the experience. Therefore, the product was not differentiated in a way that motivated consumers to continue to repeat purchase. The product did not help the brand achieve a positive, sustainable emotional connection to consumers. It ultimately failed to fulfill an important role in the lives of the target consumer. Marketing research did not help guide successful development of the product.

---

3. Olivia Putnal, "The Biggest Food Product Duds of All Time," posted April 23, 2009 from WomansDay.com. http://www.womansday.com/Articles/Food/The-Biggest-Food-Product-Duds-of-All-Time.html

It is important that companies learn from their mistakes. The continued success of Clearly Canadian after the Orbitz failure suggests the company did learn from its mistake. By ensuring that innovation is founded, directed and differentiated, the benefits from success can be realized. Successful innovations benefit companies financially by achieving or exceeding revenue forecasts and profits. Innovation can also provide excitement in the marketplace, creating new enthusiasm and value for whole product categories. This excitement can also extend to employees and staff and, in turn, that excitement can inspire an increase in the number of new ideas.

As seen with the case of Orbitz, however, failure can be a painful lesson. Undirected, unfounded or undifferentiated innovation wastes valuable company assets. Failure discourages employees and, in extreme cases, can lead to company failure.

Voice of the Consumer

The success of Frito-Lay's Baked! Lay's and failure of Clearly Canadian's Orbitz are the book-ends of innovation case studies from the 1990s. This was a time of transition, when many companies in the food and beverage industry began to realize the importance of the voice of the consumer to successful innovation. This time of transition also saw the development of a number of innovation processes (e.g. Stage-Gate®) and advances in the techniques to bring the voice of the consumer into the beginning (i.e. "fuzzy front end"), middle (i.e. "concept development") and end (i.e. "product guidance") stages of the innovation process. These techniques typically involved face-to-face marketing research and sensory research methodology.

In the early years of the 21st century, there was a dramatic shift in the marketing research industry from face-to-face ("offline") research to Internet ("online") research. This shift in research methodology dramatically decreased the cost of the voice of the consumer. As a consequence, innovators were able to more rapidly access more data than ever before to make decisions throughout the innovation process. However, the shift to online changed the mode of listening to the voice of the consumer. The "vibe" that researchers were able to get from face-to-face research was replaced by rational responses in a typical Q&A self-reported mode of research. While less expensive and faster to attain, rational-based information is subject to many more errors in judgment and bias. Further, the rapid popularity in online panels brought with it issues in data quality. As a consequence, the ability to hear the true voice of the consumer has been weakened.

These changes in how companies have been bringing the voice of the consumer into their innovation process are extremely important to note. As quantitative research increased in popularity, the quality of the information content decreased, resulting in fewer insights into the "whys" of consumer behavior, thus compromising the ability of companies to establish a solid foundation for innovation, to be able to accurately direct innovation and to bring to market products that are differentiated in ways that achieve positive emotional impact.

Today, companies are realizing that more information does not mean better insights; they acknowledge that they need to deepen insights through research – not only to listen, but to observe and dialogue in new ways. The number of new research techniques being launched today is unprecedented in the history of marketing research. Companies are realizing that what they need is not more information, but better information. The research community is grappling with the question, "what is better information?" A better question might be, what information can the organization best use to achieve innovation success?

The P&G Success Story

In 2000, Procter & Gamble (P&G) was in disarray. The company was sinking under the weight of too many new products and organizational changes. Only 35% of their new products met financial objectives. R&D productivity had leveled off with R&D spending unable to meet growth objectives. New CEO A.G. Lafley was brought in to turn things around. Lafley knew what innovation meant to successfully grow a company.

In his book *The Game-Changer*,[4] Lafley and co-author Ram Charan write about a vision for the "innovation company." This vision started with a basic premise that the consumer was at the center of everything. This led to a consumer-centric definition of innovation: a creative process, expressing how consumers view and experience the product. They further defined the consumer experience in broad terms: the brand, the product function and other types of benefits, the business model, the supply chain, and the cost structure that makes the product successful. They also changed the P&G organization in a number of ways to achieve the following objectives:

- Make sustainable organic growth through innovation a priority.
- Think about innovation in different ways.
- Increase the innovation pipeline through co-innovation.
- Better direct innovation through research that achieves learning for how consumers want to use P&G products.
- Organize around innovation.

Lafley further restructured the P&G innovation process – simplifying it into the following steps:

1. Understand the customer.
2. Establish a product strategy.

---

4. A.G. Lafley and R. Charan, The Game-Changer: How You Can Drive Revenue and Profit Growth with Innovation. Random House, Inc., New York, NY, 2008, pp. 336

3. Use a simple process to gather and convert ideas into concepts and prototypes.

4. Apply a simple development and qualification process.

In an interview in 2008, Lafley discussed his realization that P&G had to go to consumers to find out how they were using, and wanted to use, P&G products.[5] He placed special importance on developing internal consumer research and design functions. The consumer research at P&G during this time became more experiential, putting a new emphasis on observing consumers in the context of real-life experiences. In addition, he realized that the 7,500 internal researchers in the company were not sufficient resources to create the level of innovation required to turn the company around. So, he set a goal calling for 50% of all innovation to come from *outside* the company. He embraced the concept of "open innovation," establishing leading business processes to broaden the company's innovation resources to 1.5 million researchers and scientists worldwide.[6]

By the time he stepped down to become full-time Chairman for P&G in June 2009, their stock had increased from $28 to $52 per share. Lafley had not only galvanized the company around innovation, he made sure innovation was focused in the right ways. He had pushed P&G toward higher-margin areas like health, beauty, and personal care with $23 billion in brands, including Tide, Crest, Pampers, Gillette, Olay, Pantene, and the latest addition, Gain laundry detergent. At the end of his tenure, profits had tripled to more than $10 billion on $76.5 billion in revenues.

Yet, in spite of all of Lafley's vision and successes, there are indications that he felt the job was far from complete. *Business Week*'s, Bruce Nussbaum, interviewed Lafley several times over his tenure at P&G. He wrote the following in his column,

"Even with those abilities, Lafley believes that the job of changing P&G's culture from conventional to innovative is only 10% accomplished – at best. Building an innovative culture in old-style, big organizations takes a generation, even with the best of leadership. Corporations in the US especially, have not had Lafley-like CEOs and it's one of the basic reasons why it is in economic and geo-political decline."[7]

Lafley realized what so many companies do not – that successful innovation requires more than simply better information or better processes. Innovation needs to be integrated into the very DNA of a company. Innovation must also be

5. A.G. Lafley, Innovating at Procter and Gamble, Front End of Innovation (FEI), April 2008. http://frontendofinnovation.blogspot.com/2008/04/cnbc-has-recently-posted-interview-with.html

6. L. Huston and N. Sakkab, Connect and Develop: Inside Procter & Gamble's New Model for Innovation, Harvard Business Review, March 2006, 58–66

7. CEO Legacies: A.G. Lafley vs. Bob Nardelli, http://www.businessweek.com/innovate/NussbaumOnDesign/archives/2009/06/ceo_legacies_ag.html

driven by strategy and supported by culture. P&G was ahead of the curve in many ways compared with its competition, yet its culture and strategy were still lacking and, as it turned out, they were not able to adjust to rapid consumer change in behavior.

INNOVATION IN THE FACE OF CONSUMER CHANGE

Lessons from the 2008–2009 Recession

The recession of 2008–2009 shook the world economy to its core. Great innovation companies such as P&G felt the impact of consumer change. The issue was not so much that consumers changed, but that companies could not react fast enough.

For the first time in years, P&G saw its growth rate fall below double digits. Consumer reactions were demonstrated clearly with Tide, P&G's flagship brand. Tide consumers talked with their pocketbooks:

- 19% reported trading down to value brands because of recession.
- 81% indicated they were likely to keep buying value brands after the economy improves.
- Tide's market share in liquid detergent dropped to 42.7%, a 2.4 percentage point loss in the twelve weeks that ended April 18, 2009 in food, drug and non-Wal-Mart mass outlets.[8]

Yet, the recession was only a symptom of a much bigger and more impactful trend. The speed and severity of the economic impact on the consumer packaged goods industry was directly related to the emergence of a much more engaged, demanding and emotionally-driven population. Peer-to-peer networking spread the bad financial news. As more and more bad news built, consumer confidence dropped at a speed never before observed. Not only were billions of dollars lost from property investments, but trillions of dollars were lost through the change in consumer behavior.

The food and other consumer product industries suffered greatly throughout the recession because they did not understand the dramatic impact social networking was having on consumer behavior. The culture, strategy and processes of the food industry moved like a dinosaur in the face of a dramatic acceleration in consumer behavior change. Their innovation process was consumer-driven, but it was not behavior-driven.

Human Nature

Human nature had a lot to do with the unprecedented speed of consumer change in response to the events that triggered the 2008 recession. People were

8. Why We Will Remember Tide Thursday, http://consumeredgeresearch.com/news/item/why_ we_will_remember_tide_thursday/

attracted to social networking because it fit with their true nature – a deep desire to interact with other human beings. The development of social networking media in the mid-2000s enabled people to act out their true nature en masse, and so emerged a much more reactive society.

The year before the recession, Mark Earls published a book entitled *Herd: How to Change Mass Behavior by Harnessing our True Nature.*[9] Earls characterizes the true nature of humans as "herding creatures," and stresses the importance of applying an understanding of consumer mass behavior in terms of peer-to-peer interaction. He suggests that we should redefine the concept of a market from a classical definition of individuals and their attitudes and behaviors to a more contemporary definition of peer-to-peer interactions and how they impact market behavior. He stresses that it is only through the understanding of peer-to-peer interactions that we can start to understand how relationships influence consumer behavior.

A number of other thought leaders are beginning to add voice to how to rethink innovation in this new world. Important contributions are emerging from the field of social psychology. If markets are to be defined in terms of peer-to-peer interactions, then the traditional model of mass marketing no longer applies. If influence is now in the hands of consumer peers, then how do companies dialogue effectively to discover how to focus innovation? If the power has now shifted to the prosumer – that consumer specifically focused on seeking and sharing information on products with others – then how must innovation change to effectively inspire consumers to behave in ways that have long-term positive economic impact for food companies?

Additional thought leadership is also emerging from the field of behavioral economics. This field of study is being applied to gain new insights into how consumer perceptions, emotions and social dynamics interrelate to drive market economics. Its economic theories are being touted as a replacement for neo-classical supply and demand economics. The coalescing of this knowledge into a framework empowers the innovation team with new theories that can be applied to solve the innovation problem in the face of change. The capability to identify emotional drivers is a huge step for companies on their path to becoming more behavior-driven. By applying perceptual and behavioral psychology, companies are better able to link the sensory qualities – perceived consciously or unconsciously – to their innovation decisions, enabling innovation to be focused on delivering experiences that drive behaviors.

These new thoughts are contributing to change in innovation strategy. Food companies now have to align corporate behavior with the values of their consumers; they have to embrace the new world order of consumer behavior. To do this, companies must demonstrate social responsibility and global citizenship to align with consumer expectations as to how companies should behave.

---

9. M. Earls, Herd: How to Change Mass Behaviour by Harnessing Our True Nature. John Wiley & Sons, Chichester, 2007, pp. 339

The application of these fields of psychology provides business strategists and managers with a better understanding of how to achieve sustainable business practices in spite of change. Of utmost importance is the building and maintaining of consumer trust. Companies must therefore not only focus their innovation on messaging that leads to initial trial behavior, but also on how consumers' experiences align with their expectations. This requires that the innovation and development processes take a much more holistic approach.

Innovation in the face of consumer change also requires a new approach to research, starting with the capability to ask the right questions and to deliver better insights. Perceptual and behavioral psychology are also enabling a new era in the delivery of better insights – insights that are behavior-oriented – deepening understanding into the "whys" of consumer behavior. This includes a number of new ways to listen to peer-to-peer dialogues between consumers and to engage consumers in dialogue. These research methods and techniques are beginning to uncover new ways to identify which emotions are tugging at the hearts of people and how to react to those emotions. Companies that embrace these new techniques and methods gain advantage by being able to imagine how to create a world of brands and products that calms consumer fears and delivers what is hoped for.

Behavioral Insight

It is through the application of these two fields of psychology that innovators achieve behavioral insights – knowledge into how to focus innovation to make it truly behavior-driven. *Behavioral insight applies the right science to answer the right questions.*

As an example, if you are a manufacturer of jelly beans and seeking insights into how to achieve better quality – then you need to apply the science of perception to understand how to produce jelly beans of more consistent quality. If, however, you are seeking to develop a new type of candy, the science of pleasure might be applied to dictate how much sugar is optimal to dial up the emotive connection. However, these insights are not behavioral insights. These insights are more focused on consumer perceptions than behaviors. They are examples of insights into measures of product affect, i.e. intrinsic properties of products. These are measures of quality and pleasure. Quality and pleasure may not always translate directly into behaviors.

Since all candy with high sugar content is fundamentally pleasurable, more sugar will not necessarily differentiate your product. This alone will not necessarily create emotional connections to a brand. What you need is a framework that goes beyond just making a product pleasurable and consistent in quality. You need a framework that helps you understand how to differentiate your product in ways that achieve emotional connections with consumers. This framework uses the science of psychology to understand people on an emotional level. It helps innovators understand how to differentiate products in ways beyond functionality, sensory qualities and hedonics (i.e. pleasure). It helps innovators also differentiate products on a social and emotional basis.

Pop Rocks® are a great example of a strategy for development that tapped into more than just sensory pleasure. The candy contains tiny air pockets of carbonation that are released as it melts in your mouth, creating a mild "crackling" sensation and "popping" noise. This product provides more than sensory pleasure. It impacts consumers emotionally. The product can elicit surprise at the unexpected first time consumption experience. The product can elicit enjoyment during repeat consumption experiences. The product can provide for a social experience – eliciting intrigue when the product is first introduced by a friend.

The importance of understanding how a product impacts consumers on an emotional basis is essential to the innovation and development process. Why? Emotions are the basis for most consumer behaviors. Having a usable framework that links the product decisions to emotions enables innovators – marketers, designers and developers – to rapidly focus innovation in ways that increase the chance for success. Insights from research become behavioral insights when they link product decision to the elicitation of emotions that drive consumer behaviors.

THE INNOVATION REVOLUTION

The points made in this chapter about how innovation has been changing over the past 20 years can be summarized as an evolutionary process. Companies such as Frito-Lay and P&G have evolved from a technology-driven (i.e. "build it and they will come") to a consumer-driven innovation company. They have been leaders in the development of innovation – using the voice of the consumer to focus innovation and to use innovation as a key pillar of business strategy. However, the emergence of new means of peer-to-peer social networking has changed everything. Being consumer-driven in your approach to innovation is no longer enough. In a short time period, people have fundamentally changed how they interact socially, what sources of information they trust, and how they behave as consumers. These changes require that companies undergo more than just an evolution; they must undergo a revolution.

Industry change is not an easy task – especially when it comes to changing culture. A.G. Lafley believed his efforts to change P&G innovation culture were only 10% realized. To make innovation a key pillar of business culture, innovation must become embedded into the whole business model, i.e. impacting a business's strategy, its operating structures and infrastructure, such as information systems.

The starting point to making this change is raising the bar in how behavioral insights through research are attained. Once this bar has been raised, researchers will find new vistas for applying those behavioral insights – converting them into knowledge and applying that knowledge throughout the innovation and development process.

Behavior-Driven Innovation

If Pop Rocks were invented today, what behavioral insights might be generated and used by marketers, designers, and developers to achieve rapid, successful innovation? A more behavioral approach to innovation starts with research that helps innovators imagine what consumer experiences might lead to product differentiation on an emotional basis. Perhaps consumers are seeking exciting, new candy experiences that will surprise them. Perhaps consumers are seeking social experiences that elicit intrigue from peer-to-peer interactions. Starting with an identified set of emotions and behavioral outcomes as a target, you can start to see the possibilities for a fundamental shift in innovation. The result of this shift is a new focus on achieving behavioral outcomes through innovation. This focus enables innovators to become inspired in a new way, opening up more degrees of freedom for how to create emotions such as surprise or intrigue. This focus also leads to new thinking about how to guide decision making throughout the innovation and development process.

Behavior-Driven Innovation (BDI) is an approach that starts with the behavioral outcome in mind and maintains that focus throughout the innovation and development process. It has the capability to impact whole organizations, due to the breadth of different disciplines that are typically involved in product innovation and development. In this way, BDI is a platform that can be used to create a culture of innovation – as envisioned by Lafley.

Raising the Bar

Consider the traditional usage and attitudinal research study. According to ESOMAR, $318 million was spent in 2008 on traditional online survey usage and attitudinal (U&A) studies within the United States. These studies have no means to provide behavioral insights. They can help track what consumers are doing, yet are dependent on rational-based survey responses to identify consumer attitudes. By interrelating attitudinal information with self-reported behavioral claims, companies hope to achieve behavioral insights into why consumers are behaving as they do. According to behavioral psychologists, this approach to achieving behavioral insights simply does not work. Something is fundamentally missing in the richness of information generated from a typical U&A study.

A behavioral approach to innovation begins to raise the bar by generating new information through research that can be processed into behavioral insights. This new information fills a knowledge gap for how current or envisioned experiences lead to emotional impact – motivating consumer behavior. It fills knowledge gaps for what kind of brand relationships consumers are seeking. And, it fills knowledge gaps for why qualities of a brand and/or product elicit specific emotions – positive and/or negative – about an experience. It goes beyond the learning from a traditional U&A study.

Advancing research methods to generate behavioral insights will start you on the pathway to the innovation revolution. However, the journey will still be incomplete. Your organization must be able to utilize behavioral insights. The capability to generate behavioral insights leads naturally to a shift from a linear, inflexible process to a more flexible and adaptable one. It requires that innovation teams work more collaboratively, using their different vantage points to generate insights by actively participating in research that engages with the consumer. These shifts in process finally lead to a change in culture – a culture that blurs the traditional roles of domain experts working in silos of specific aspects of the consumer experience. It leads to a cultural shift where innovation and development is holistic.

An Evolving Path Forward

Once the first step is taken to becoming behavior-driven in the way research information is generated and processed into knowledge that informs innovation, the path forward leads to a cultural paradigm shift. This shift will impact a company's infrastructure for information processing, strategy development for business units and brands, and structural organization of how people work together. Fully implemented, BDI has the capability to complete the remaining 90% of the process (as envisioned by Lafley) in transforming companies from the "build it and they will come" mentality of the 20th century into the behavior-driven company of the 21st century.

Key Points

- Unfounded, undirected, or undifferentiated innovations are the three primary reasons for product failure in the marketplace.
- The research shift from off-line to on-line in the early 2000s has weakened the ability of companies to listen to the true voice of the consumer.
- The P&G success story in the early 2000s was driven by the vision of A.G. Lafley to focus the organization on being an "innovation company." Their success was attributed to innovations that were founded (i.e. strategic choices were made to play in markets with higher margins), directed (i.e. driven by the consumer), and differentiated (i.e. applying an open, simple four-step disruptive innovation process).
- The pace of consumer change has accelerated in the 2000s, requiring that even successful companies rethink their approach to innovation.

- Innovation based upon behavioral economic theory can be achieved by applying the science of perceptual, behavioral, and social psychology to generate behavioral insights. These types of insights enable the inspiration of divergent thought and the guiding of convergent thought throughout the innovation process.
- Behavior-driven innovation (BDI) is a business framework for sustained business growth through innovation. It involves the generation of behavioral insights that inspire and guide innovation, impacting the DNA of business culture and the development of corporate strategy, business processes, and information systems.

Chapter 3

The Innovation Team

When all think alike, then no one is thinking.

Walter Lippman

INNOVATING AS AN ACT OF COLLABORATION

The above quote by two-time Pulitzer Prize winner, Walter Lippman, emphasizes the power of critical thinking through collaboration. Different perspectives cultivate opportunity for breakthrough ideas to emerge. Breakthrough ideas may be driven by market insights or an R&D-driven invention looking for an application. Yet, the spark for innovation often emerges from an unexpected source – a consumer comment or expressed thought during a discussion where very different perspectives are shared among members of an innovation team.

This chapter will explore the power of diversity of thought and how different disciplines contribute to divergent and convergent thinking. The roles within the innovation team will be defined, casting a broad, multidisciplinary net to capture thought diversity from not only professionals working within a company, but also outside as business partners. This broad definition will also consider how technology is shaping the role of the consumer in contributing to innovation. This will lead into a discussion of the innovation process itself – how the uncertainty of where an innovation spark will come from necessitates what the innovation effort should foster, rather than force innovation through a rigid process.

THE INNOVATION TEAM

Innovation teams function best when they are comprised of a melting pot of different perspectives. The diversity of perspectives comes from the mix of professionals trained in, and having experience in, different disciplines that contribute to thought diversity. This section will characterize innovation teams as comprised of four types of contributors: Innovation Manager, Innovator, Sensory Researcher, and Marketing Researcher (see Figure 3.1).

These contributors bring thought diversity to the innovation team in a number of ways. The schematic in Figure 3.1 portrays the multidisciplinary

Breakthrough Food Product Innovation Through Emotions Research. DOI: 10.1016/B978-0-12-387712-3.00003-7

FIGURE 3.1   The multidisciplinary nature of innovation within the innovation team. (Please refer to color plate section)

diversity of thought from these contributors to the innovation team. On the left are the disciplines that contribute to building team knowledge through research. On the right are the disciplines that apply knowledge for innovation. Likewise, the upper disciplines contribute to building and using knowledge to develop innovation strategy, while the lower disciplines contribute to building and using knowledge for product development.

Team Members

The strongest innovation teams in the food industry have members who are able to collectively contribute their respective knowledge from this array of disciplines. Team members may have deep domain expertise from one discipline or have broad professional training and experience in multiple disciplines. Team members can come from inside their organization, or from outside the organization as co-innovation partners or as co-creating consumers.

Innovation managers contribute their knowledge to the team by owning the development of innovation strategy. They contribute to the team by serving as the primary client (i.e. the brand or product line management), allocating resources for the team (i.e. portfolio management), serving as the team lead (i.e. managing how the team is organized), or managing co-innovation (i.e. how ideas flow into the team from contributors outside the team). *Innovators* apply their domain of expertise to create and to make in-process marketing, design and development decisions within their team roles. *Researchers* generate insights from research that inspire creativity and guide decision making.

Researchers come in two flavors. *Sensory researchers* are responsible for generating insight into how a product impacts the consumer experience (perceptually and emotionally). The methods and techniques used by professionals in this role tend to be based upon the disciplines of perceptual psychology and consumer behavioral psychology. *Marketing researchers* are responsible for generating insight into how markets are impacted by innovation. The methods and techniques used by these professionals are typically based upon the disciplines of behavioral economics (e.g. choice behavior), social psychology (e.g. dispersion behavior), cultural anthropology (e.g. cultural norms), and the emerging field of digital media and marketing (e.g. using digital consumer and market information).

Working Together as a Team

In many ways, innovation teams are analogous to football teams. The innovation manager – like a coach – is responsible for building the team, and developing and implementing a winning strategy. In the case of innovation, the innovation manager is responsible for developing the innovation strategy – i.e. how to use the strengths of team members in their respective innovation roles to achieve business goals.

Innovators – like receivers and ball handlers – are members of the team responsible for implementing the innovation strategy. They score touchdowns by creating space for the innovation team to operate – discovering opportunities for new and improved products, scoping their conceptual designs, designing product requirements, and developing products. They tend to use their intuitions – seeking out that innovation spark – to create new possibilities.

The researcher is very much like a quarterback. While innovators and innovation managers might be superstars in their respective role, it is the researcher that is often the difference between winning or losing – i.e. the researcher is often the game changer. Quarterbacks must work with both the coach to develop the game plan and with the receivers and other ball handlers to execute the game plan. They must have skills to get the ball to the hands of receivers and ball handlers – i.e. being able to generate and communicate insights.

It is the researcher that has the power to raise the innovation bar by delivering insights that link a holistic definition of the product to the experience for the consumer. When researchers are able to deliver these types of insights, they change the game for both the strategist and innovator. Insights that answer questions about why consumers are changing purchase behavior enable innovation managers to establish strategy to adapt to that change. Insights that answer questions about why consumers respond to media mix or positioning enable marketers to better adapt their mix or messaging.

Insights that answer questions about why product and package features, functionality and sensory qualities elicit emotions enable designers and developers to adapt.

The Sensory Research Function

Sensory science was founded in the 1950s as an application of perceptual psychology to measure properties of foods using human subjects. These applications were mostly in the laboratory. However, in the 1980s, the field began to shift its application to measuring not only sensory perceptions (e.g. recognition, characterization and rating intensity of perceived sensations), but also sensory affect such as "liking" (i.e. sensory pleasure) and preference with consumers. This put sensory into heated conflicts with marketing research. Today these "turf wars" have, to a large extent, subsided; however, the boundaries remain unclear – especially when it comes to research that involves consumers evaluating a mix of product concepts, packaging and packaged foods.

Certainly the expertise of the sensory field has been, and remains, in its methods and techniques to measure how people perceive products through their senses. Sensory research has traditionally applied this expertise in support of the development of food products. The methods it employs tend to be very scientific, taking a reductionist approach to generate consumer product insights. However, this appears to be changing. There is a growing awareness that product developers need inspiration and guidance based on measures that go beyond liking – to capture measures that lead to insights into how products impact consumers on an emotional basis.

Insight into emotions requires a more holistic approach to research. This means that sensory research is now researching products that are fully branded and in home-use situations. Sensory research now includes choice measurement. The boundaries appear to be even more blurred than ever.

What appears to be emerging is a new distinction between sensory and marketing research. The distinction is shifting to one based on domain expertise. In many companies, sensory research is now responsible for all consumer product insights, marketing research for consumer insights.

In many ways, this redefinition of roles makes perfect sense. As both sensory researcher and marketing researcher sit on innovation teams, the question for role play is not so much about whom they support, but what type of insights they provide. In this case, the sensory researcher is able to play the role of chief provider of consumer product insights provided a broader, more holistic approach is taken to understand emotional impact to products. The problem is that many sensory researchers have yet to transition from a reductionist approach that limits their capability to delve into emotions research. However, emotions research is undoubtedly the hottest topic in the sensory field. So, change is coming.

The Marketing Research Function

The primary innovation role of the marketing researcher is also changing. Traditionally, the marketing researcher has been focused to support the marketing and business management functions. The research methods used have their roots in the disciplines of social psychology (e.g. research methods to understand market dynamics), cultural anthropology (e.g. ethnography) and digital media and marketing (e.g. tracking product and brand performance).

Marketing researchers apply their methods and techniques to deliver consumer insights into how consumers are changing in their relationships to brands. These insights build knowledge into what consumers seek, how they share what they have found, and what they select. However, it appears that insights delivered by marketing researcher staff often fail to be of strategic value to innovation managers.

In a 2009 survey by the Boston Consulting Group, 819 product line managers and consumer insights staff working for consumer packaged goods manufacturers responded about their respective perceived value of consumer insights.[1] The report found lower than expected agreement by business managers that consumer insight personnel "consistently answer the question *'so what'* ..." (34%); "understand the business issues as thoroughly as line management" (41%); and "translate research findings into clear business recommendations" (32%). These results indicate a need for marketing researchers to deepen their insights to be of more strategic value to innovation teams.

It may be that marketing researchers simply do not have sufficient time and resources to uncover insights into the "whys" of consumer behavior. Survey results such as this are strong indications that change is in the wind for the marketing research community as well. Emotions have also crept into the marketing research vernacular. The emerging discipline of behavioral economics has led to awareness that emotions often are stronger motivators of consumer behavior than rational thought. The understanding of why consumers connect (or disconnect) with brands is largely a question of emotional impact. The seeking behaviors that drive prosumers to take action are emotions-based. The desire to share experiences with peers is emotional. The deepening of consumer insights, through an understanding of how emotions play a role in behavior, lead to a path that goes beyond what can be attained from traditional marketing research methods.

The innovation manager who is reliant on consumer insights to set innovation strategy requires insights that go beyond questions of "how" consumers

---

1. M. Egan, K. Manfred, I. Bascle, E. Huet and S. Marcil, The Consumer's Voice – Can Your Company Hear It? Boston Consulting Group, Center for Consumer Insight Benchmarking, November 2009, pp. 31.

are changing and "what" they are doing. They need to also understand "why." To be sure, change is coming here as well.

Managing the Innovation Function

Innovation managers play a diverse set of innovation roles. Some managers are on the innovation team – perhaps serving as a team lead. Others are not on the team, but play an important role as the primary client. This is often the case in the food industry where brand or product (line) managers contribute to the team by being the primary client. In other cases, managers contribute to the innovation team by allocating resources, or building structure for how the team will operate within the organization. Examples of this wider role are professions not on the team, but serving the innovation team by managing the company's portfolio of innovation projects or being responsible for managing the innovation and development process itself. Whether on the team, or off the team in a support role, these innovation managers play important roles in developing innovation strategy.

Innovation strategy sets the boundaries for innovation that focus the innovation team. These boundaries include key constraints to where the team might discover the biggest, most attainable opportunity with the right risk–reward. These boundaries might be driven by technology (e.g. research and development hurdles), market factors (e.g. consumer targets, competition, or use contexts), or business factors (e.g. business focus, risk–reward tradeoffs, investment capital, supply chain issues).

The mapping of these boundaries requires a solid understanding of the company's goals and how those goals translate into goals for each brand they produce. Once a brand knows where it is going (i.e. has developed a brand strategy), it is the responsibility of the innovation manager to ensure that innovation teams have a strategy to act upon. In this way, innovation managers can be viewed as being responsible for the front end of the innovation process.

Brand strategy sets the focus on how to move the brand from where it is currently in the minds of consumers, to a future state. This understanding, in the form of a brand strategy, leads naturally into decisions for whether innovation will result in a new product, a new product line or a product line extension – i.e. how will the brand strategy translate to expand the portfolio of products under the brand. Will the brand become elevated to be an endorser of other brands? Will the brand seek a defensive or offensive strategy to win against its competition? These questions are essential in developing a comprehensive innovation strategy.

It is the innovation manager, typically in the role as brand manager, who is often responsible for making innovation strategy decisions. However, the innovation manager cannot lead without support from the rest of the team. The marketing researcher must be able to provide consumer feedback about the "whats" and "whys" for how the brand and consumer relate. From these insights, the innovation manager is able to lead the innovation team in the development of strategy for the brand, the brand's portfolio of products and for the front end of the development process.

Consider the case study of Kettle Foods, a natural-snack producer out of Salem, Oregon. The Kettle Foods brand is targeted to support a community of consumers interested in consuming organic snack foods. In 2004, Kettle Foods turned over their product development to fans inviting chip lovers to help select new flavors by voting for their favorites.[2] The voters were offered five "hot" nominees, including Wicked Hot Sauce, Mango Chilli, Jalapeno Salsa Fresca, Orange Ginger Wasabi and Death Valley Chipotle. Death Valley Chipotle was the winner, receiving the largest number of votes. By involving the community in the discovery and scoping phases of innovation, Kettle Foods developed products "designed for their community." In addition, they showed commitment to that community by publicly stating support from their "employees, their craft, and their community" for their lifestyles and values. The authenticity of their support was also demonstrated by financially matching contributions to the Death Valley Natural History Association through the corporate website that markets Death Valley Chipotle™.

This is an impressive innovative case study showing how marketing research can become of strategic value to innovation managers and innovators. By engaging core target consumers in this way, the brand not only gained insight into an opportunity to launch a new product line, the brand was able to demonstrate its authentic support to its loyal consumer base. Tactics for innovation strategy development such as this are discussed in more detail in Chapter 5.

The Innovator Function

Innovators are domain experts that serve a common function within the innovation team – to use research information and their domain knowledge to take an idea through the innovation process to launch. They are the marketer, designer, creative chef, package developer and product developer. Due to their professional diversity, they may be organized within different corporate functions such as research and development, marketing, innovation, and product development.

Innovators need inspiration to focus their creativity and expertise. Their inspiration comes from consumer and consumer product insights gained through consumer engagement and from collaborations with other innovation team members. These professionals also require constant feedback from consumers to guide their respective decisions within roles.

---

2. Press Release: Kettle™ Brand Potato Chip Fans Bring Death Valley Chipotle™ to Life. Newest People's Choice Flavor to Spice Up Store Shelves, Salem, Ore., January 14, 2008

Innovators need collaboration to do their job. The messaging, product and package design all interact too much to be developed in silos. While one might try to develop a product in pieces and parts, the consumer does not experience the product this way. To the consumer, products are experienced perceptually and emotionally as a whole, complete product, leading to feelings that build, reinforce or tear down the brand–consumer relationship. It is through the product development steps where team collaboration is most important. The brand, product, packaging and messaging all must come together to form a consistent, unified product experience for consumers.

Innovators play chief roles throughout the product development phase of the innovation process. They take the developed innovation strategy, and implement it. In the early phases, the team must be able to discover the product experience that leads to the biggest market opportunity. It is here where the creative inspiration for the whole team is key in identifying the target consumer, usage cases, and elements of the experience that differentiate the product from alternative choices. The marketer must be able to contribute their understanding of why a consumer might find a given experience valuable. Designers must be able to contribute their understanding as to how the context of the usage case might interact with the product. Developers must be able to contribute to the practical aspects of how the components of the product and package interact, in a given context, to form a differentiating experience.

Consumer product insights are critical in building the knowledge of innovators. Each innovator may glean different nuggets of knowledge from every insight. An example of how consumer insights drive innovation is the case study for the development of an "on-the-go" food product for a major US food company.[3]

The innovation team was comprised of a product developer, packaging designer, marketer, sensory researcher, consumer insights researcher working within the company, the marketing research supplier, and a team of "creatives" from the food company's advertising agency. The goal for the innovation team was to develop a new sub-brand for a popular meal entrée brand – extending the brand from a stovetop-only offering into something that could be quickly prepared and consumed "on-the-go." The team was extremely excited about a concept that had shown great promise as being differentiated for its "on-the-go" experience. The team had only a six-month window to bring the product to market.

The task at hand was to design a prototype, translating the concept into a product name, package, and formulation. In addition, the flavor and texture profile had to fit the promise of the brand. The sensory cues from the packaging and visual appearance of the product had to elicit positive emotional impact.

This group focused their expertise and creativity to generate several potential names, three alternative packaging prototypes, and five alternative product formulations. In addition, the team set up sessions to engage brand

---

3. InsightsNow (unpublished data)

loyal consumers as research participants. In each session, a different group of 30 participants assessed packaging alternatives, tasted the different formulations and evaluated alternative names against the concept.

The sensory researcher and marketing research partner were able to analyze quantitative results at the end of each session. This provided insights into what packaging, product formulas and names were preferred. In addition, the marketing research partner provided real time profiles from responses to the packaging, product formulations and names. These profiles were used to select several participants for one-on-one, co-creative sessions with innovators to determine why they responded as they did.

The result was an interactive co-creation session focused on how the design of respective components of the product (i.e. package, formula, and name) impacts the consumer experience. In the end, a small group of 15 participants were allowed to take home a prototype. These participants were invited to blog together about their on-the-go preparation and consumption experience contexts.

The results also included insights into modifying the packaging design. In one of the early sessions, insights emerged into how a particular ingredient might be changed to improve the formulation as well as an improved product name. While the formulation change could not be tested in the later sessions, the new name could. This led to an increase in learning, as the new name was able to be validated with later session research participants. This dynamic approach to research is an example of how innovation teams can speed up their learning. In this case, the product launch was not only on time, but one of the most successful in the history of the company.

This case study shows the power of collaboration. The innovation team was able to engage with consumers not only to guide the selection of package and formulations, but was also inspired to create a new name. Innovators were able to rapidly co-create, focusing divergent thinking on how the design elements of the product interacted with consumers in specific use contexts. The company was able to quickly translate the concept into a viable prototype that was rapidly developed into a successful commercial product.

HOW DO INNOVATION TEAMS LEARN?

Traditionally, sensory research, and to a lesser degree marketing research, have applied a "scientific approach" to generate insights for innovation teams. This approach works well as a learning mechanism for scientific endeavors involving quantitative research. However, this type of learning does not fit with the collaborative learning processes used within innovation teams. This type of learning is more qualitative – based on observations and intuitions. An alternative approach that fits better with the collaborative nature of learning within innovation teams – learning through research that involves a mix of quantitative and qualitative methods – is shown in Figure 3.2, which depicts a schematic termed the "Learning Cake."

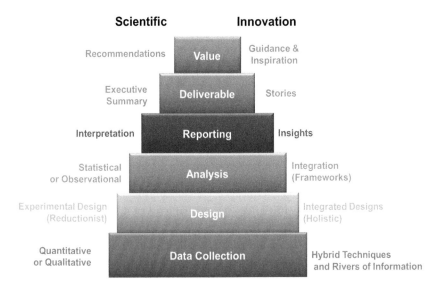

FIGURE 3.2   The Learning Cake – a comparison of two approaches for innovation team learning. On the left is a traditional (scientific) approach. On the right is a more contemporary integration approach. (Please refer to color plate section)

The scientific approach (shown on the left) starts with a base of quantitative or qualitative information that is generated through an experimental design. This reductionist approach has traditionally been the approach of choice for sensory researchers to generate insights. The analysis is typically statistical for quantitative data and observational for qualitative data, resulting in the issuing of a report with interpretations. An Executive Summary is often provided as a snapshot of the key findings. This leads to recommendations for the next action step in an iterative process that follows the scientific method (i.e. hypothesize → test → learn → hypothesize → … etc.).

Today, a new learning process is emerging for innovation. This process includes continuously dipping into rivers of consumer information and/or capturing innovative mixes of quantitative and qualitative data through research design that is much more integrated and holistic. These designs establish the basis for emotions research. In place of statistical analyses, methods are emerging for the integration of quantitative and qualitative research. This process of integration is leading to new types of insights that are more emotions-based. Instead of interpretive reports with key learning communicated through an Executive Summary, this new approach extends learning by building stories that make insights real and tangible. These stories are told to inspire creativity and/or to guide decision making for innovators and strategists.

The scientific method works well as an approach to build learning for scientific endeavors (e.g. inventions) that are highly quantitative and experimental in nature. However, innovation teams require a much more fluid and dynamic approach to learning. The innovation approach enables learning to be generated from insights that arise from either quantitative, qualitative, or a hybrid of quantitative and qualitative forms of consumer engagement.

Story telling is a superior mechanism for communicating insights to innovation teams comprised of individuals with diverse vantage-points and perspectives. The resulting understanding of research results through these respective diverse perspectives strengthens teams to achieve their goals.

The Data Revolution

The basis for the "Learning Cake" is data. Today, we find ourselves in the midst of a data revolution that is changing the face of innovation. The "on-the-go" case study discussed earlier is a perfect example. Using real-time software tools accessible over the Internet, the innovation team was able to instantly generate insights (on demand) into what choice alternatives were preferred. Further, the team could look through individual response profiles to select participants for an immediate one-on-one interview with an innovator. This enabled the team to accelerate not only the collection of data, but to use it to accelerate the learning process. By acquiring both quantitative and qualitative information from the same participants, the team was able to deepen consumer product insights with minimal cost increase.

Technology is also being applied to generate "rivers of information" from the Internet. Researchers are able to set up a continuous stream of "dip and learn" opportunities to generate consumer insights from digital media listening of peer-to-peer chatting within communities. These new information sources are leading to a more diverse set of commercial services to gauge consumer change (i.e. syndicated research services).

Information is also flowing from a wide array of customized research solutions to gain specific insights from open or closed communities of consumers with similar interests, behavior patterns or attitudes about brands. Online community and panel management companies are now among the fastest-growing companies in the marketing research industry (Gold, 2010).[4]

Integrated Design

It is the researcher within innovation teams who is responsible for applying information technology to increase the speed of learning at reduced costs. In a world flowing in information, researchers are often challenged to decide how best to tap into these new sources of information. Information must be organized into data that has meaning to researchers. This requires a new approach to research design – one that is much more integrated.

---

4. L. Gold, Major US MR Firms: US MR Revenue Growth Rates, Inside Research, July 2010

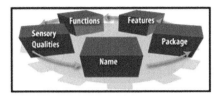
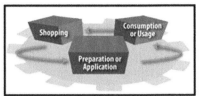
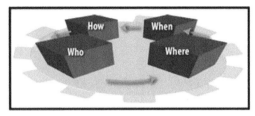

FIGURE 3.3 The three types of holistic research design: holistic product design (upper left), holistic moments-of-truth design (upper right), and holistic usage case design (lower). (Please refer to color plate section)

The traditional (scientific) approach defines data as either quantitative or qualitative in nature. The new approach for innovation defines data on the basis of how it contributes to learning. Instead of running a focus group where all data is qualitative or an Internet survey where all data is quantitative, integrated research designs mix quantitative and qualitative to gain a more holistic perspective of consumer experiences. This perspective enables not only insights to be generated into the what, which, or how aspects of consumer response behavior, but also insights into the "whys" of consumer response behavior.

Three types of integrated designs have been developed for this purpose (Figure 3.3). The first is the *holistic product design* (upper left) used to gain consumer product insights into how and why consumers react to various elements of the product experience. The second is a *holistic moments-of-truth design* (upper right) used to gain consumer product insights into how moments of truth with a product (e.g. shopping experience) impact subsequent moments of truth (e.g. preparation experience). The third is the *holistic usage case design* (lower), where consumers assess the same type of encounter (shopping, preparing or using) under different contexts. These integrated designs all enable points of engagement with research participants that are quantitative and/or qualitative.

Integration

Researchers are equally challenged to decide how to integrate these new, diverse sources of information into insights. Nonaka 1998[5] provides a key insight to

---

5. I. Nonaka, The Knowledge-Creating Company, in Harvard Business Review of Knowledge Management. Harvard Business School Press, Boston, MA, 1998

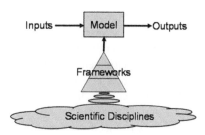

FIGURE 3.4 The incorporation of scientific knowledge into frameworks that are applied to build heuristic or statistical models that generate insights (outputs) from research information (inputs).

solving the information integration problem. He defines the term "framework" as a conceptual thought process to understand complex information.

As the cost of dipping into the information river has decreased, the complexity of that information has increased. Therefore, in order for the information flow to be used effectively, the researcher is in need of a better framework to understand the added complexity. The framework that is required is a unified set of conceptual thoughts derived from the scientific disciplines contributing to the fields of both sensory research and marketing research. As emotions are an essential contributor to both consumer and consumer product insights, a key element of such a framework would be the science of how emotions are formed.

The application of such a framework would enable quantitative and qualitative information to be equally well integrated into insights. Figure 3.4 portrays how an emotions-based framework would lead to the integration of data (inputs) for the generation of insights (outputs). The "model" would be different for quantitative or qualitative information inputs. In the case of quantitative data, the framework would lead to the building and/or applying of statistical or heuristic models with insights as the outputs. In the case of qualitative or a hybrid of quantitative and qualitative data, the framework would lead to the building and/or applying of heuristic models, also with insights as outputs. Heuristic models differ from statistical models in their reliance on thought and logic, rather than statistics to generate insights.

This book provides a roadmap to achieve breakthrough innovation by applying a new framework that is able to distill the complexity of information available within consumer insights. The new framework was developed from the knowledge amassed from the scientific disciplines underlying sensory research and marketing research. This new framework provides a process for integrating more diverse information into insights, and is further explained in Chapter 4.

Telling the Story

Integration solves a key barrier to innovation team learning. It enables both quantitative and qualitative information to contribute to insights. However, this does not solve an equally important problem in innovation – making insights relevant for the whole team. When insights are viewed as not relevant for particular team members, there is a reduction in collaboration. This leads to an assembly line mentality with product development in silos. It leads away from the more holistic, integrated approach that uses the collaborative intelligence of team diversity. This problem can be solved by story telling.

A manufacturer of microwavable side dishes was faced with gradual decline in sales for one of its product lines. An insight was discovered through a holistic moments-of-truth design that revealed the package did not look appealing. This consumer product insight was communicated to the team using a story about the shopper picking up the product and having a negative emotional reaction to how the product felt. The dilemma was the product in the package felt hard. This cued a negative emotion by not feeling like a "natural" product – a primary claim for the brand. The problem was solved by a redesign of the package and reengineering of the product to have a softer texture. If the insight had been communicated as a simple matter-of-fact – the product felt too hard – then the package designer might have discounted this issue as irrelevant to their domain of expertise, attributing the issue to product texture. By telling a story, the insight was internalized more broadly, leading to a collaboration between package designer and food engineer to solve the problem.

Inspiration

The ultimate goal of story telling is either to inspire creativity or guide decision making. Inspired creativity helps innovation teams engage in divergent thinking – to consider a broader set of alternatives. In this way divergent thinking is a team process, while creativity is an attribute of the individual. Innovation teams must be able to "think outside the box" to effectively expand the options for solving business and research problems. Therefore, it is essential that businesses focus on organizing and training teams to be most effective in being collectively creative.

Divergent thinking leads to a broader set of considerations for alternative strategies, opportunities, ideas, concepts, and prototypes. Contributors to divergent thinking include domain experts within the organization, ranging from innovation managers (i.e. strategists) to product developers and designers. Contributors may also come from outside the organization in the form of industry experts (i.e. co-innovators), or from involving highly motivated proactive consumers (i.e. co-creators).

Divergent thinking, as a component of innovation, is most effective when focused. However, focus must not stifle the creative spirit; it simply needs to have good guard rails or clearly demarcated boundaries. Focus leads to solving the right problem and having sufficient knowledge to draw upon in order to be effective. Therefore, it is essential that the front end of each phase uses the knowledge generated from the prior phase extending knowledge from past experiences through research.

Decision Making

Equally important to learning within each innovation phase is the process of convergent thinking – guided by accurate decision insights and/or metrics. Convergent thinking involves the process of narrowing alternatives, filtering out losers, and optimizing a final winner. Convergent thinking typically comes at the end of each phase of the innovation process. It is a necessary step at "phase-gates" to achieve effective portfolio management.

The chief challenge with convergent thinking is to know when you have a winner – to know when good enough is enough. For this reason, convergent thought tends to require targets, hurdles and benchmarks upon which to gauge just what is good enough. These tend to be quantitative in nature, due to the quantitative requirements of innovation managers in decision making.

Targets are goal-oriented quantities that have a historical association with success. An example is the success criterion that a product achieve 2:1 preference ratio among a target population for an alternative product. Hurdles tend to be statistical, such as a rate or rank performance against historical data. An example is a concept or product hurdle rate in the form of a "quantile" that measures how well a concept or product compares against how well similar concepts or product have performed in the past.

Many innovation teams also use benchmarks as a decision making metric. Benchmarks tend to be statistical tests that compare brands, concepts, and products against competitive or baseline brands, concepts, and products respectively. If, for example, an objective is to achieve a competitive advantage in choice against a competitive brand for a particular product, then the benchmark might be to achieve a statistical win over the competitor.

THE BEHAVIOR-DRIVEN INNOVATION PROCESS

The Eight Phases of Innovation Learning

Most companies with name brands in the food industry have a formal innovation process. Some are more structured than others. Some are very linear – with a beginning and end – some have stages and gates, others are more cyclic and iterative. Irrespective of the formal process, all effective innovation teams cycle through eight phases of innovation learning. This learning process is shown in the schematic Figure 3.5 and will be referred to as the "Behavior-Driven Innovation Process."

This process is shown conceptually as two interrelated iterative cycles. To the left is the Innovation Strategy Development Cycle; to the right, the Product

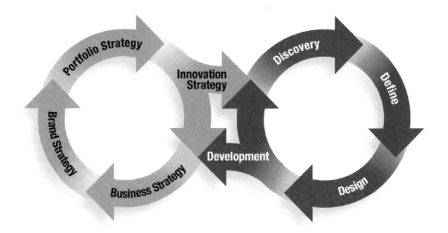

FIGURE 3.5 The Behavior-Driven Innovation Process with two interrelated learning cycles: the Innovation Strategy Development Cycle (left side) and Product Development Cycle (right side). (Please refer to color plate section)

Development Cycle. Feeding into the Product Development Cycle is the knowledge that went into the development of the Innovation Strategy. Feeding out of the Product Development Cycle is the knowledge that went into launching and tracking a new product.

The Innovation Strategy Development Cycle has four phases of learning: Business Strategy, Brand Strategy, Brand Portfolio Strategy, and finally Innovation Strategy. Each phase of learning builds upon the subsequent phase. These four phases of learning for the Innovation Strategy Development Cycle will be covered in more depth in Chapter 6.

The Product Development Cycle also has four phases of learning: Discovery, Define (i.e. conceptual design), Product Design, and Development. These four phases map well into the traditional Stage-Gate® process of Discovery, Business Case (i.e. Scoping), and Build & Validate (i.e. Design and Development). Discovery is focused on implementing the innovation strategy through research that identifies the right opportunity for a new product. This subject is covered in more depth in Chapter 7. The learning from this phase is used to focus innovation teams on defining the product through conceptual design. This is covered in more depth in Chapter 8. The subsequent learning from the Scoping phase leads naturally into the Design phase where the concept is translated into product requirements. This topic is covered in more detail in Chapter 9. Finally, the learning from building product requirements leads into the Development phase where the requirements are used to develop a final product specification. This is covered in Chapter 10.

These two interwoven cycles of learning are portrayed in Figure 3.5 as "cycles" to emphasize the importance of continuous learning within the organization. This importance cannot be overemphasized. As a food company launches a new product, the tracking of its performance against expected forecasts from earlier phases must be captured and fed into a knowledge base for sustained learning. This information helps brand teams evaluate how forecasts track against actual market statistics. This provides an opportunity for brand teams to learn about how products launched into markets actually do deliver against the set strategy to evolve their respective brands. In addition, this information helps companies consider how a particular new product launch impacts the evolution of the brand, i.e. how does it change consumer perceptions of the purpose of the brand. Further, it helps companies consider how to use product launch tracking information to assess in-process Product Development Cycle decision metrics. These topics will be covered in the final chapter, entitled *The Innovation Company.*

Within-Phase Learning Iterations

This innovation approach to learning fits with the iterative cycles that truly characterize innovation and development today. At the end of each phase, the innovation strategy development or product development activities must achieve specific objectives. "Phase-gates" are often placed within the innovation process with established hurdles or success criteria to prove projects worthy to merit additional resources. This structure helps companies manage their portfolio of innovation projects – ensuring that resources can be effectively placed against projects that have the highest likelihood to achieve the goals upon which an innovation strategy is based. This is particularly important, as the cost for development tends to increase as the innovation project nears the product launch.

All phases tend to go through a process of iterative learning involving divergent followed by convergent thinking. The insights that inspire creativity for divergent thought tend to be generated through research methods that use a mix of heuristic and statistical models. They tend to be internalized within innovation teams through the story telling process. The research methods to guide decision making for convergent thought tend to be more statistical in nature.

At every phase of the innovation process, there is a common thread of knowledge upon which new learning can be built. It is not only a base of knowledge about how consumers tend to react to brands through product experiences, but also a base of knowledge about how emotions are formed. It is the latter base of knowledge that a framework represents so that, in combination with new consumer input, the "whys" to consumer behavior can be known. This is the game changer that is missing within the food industry. The application of this knowledge leads to breakthrough innovation.

Key Points

- Innovation teams are comprised of professionals from a diverse background of disciplines. Collaboration among the sensory researcher, marketing researcher, innovation manager and innovator yields diversity in thought necessary to effective innovation.
- The researcher is analogous to the quarterback on a football team. The researcher is typically the game changer who elevates the innovation team to new levels of effectiveness.
- The role of the sensory function is rapidly changing. The fact that sensory and marketing researcher both contribute to innovation teams has led to a new distinction between sensory and marketing researcher. The sensory role is to deliver consumer product insights. The marketing researcher role is to deliver consumer insights.
- Sensory research is extending into emotions research – driven by increased awareness that emotions can play a key role in generating insights that inspire and guide product innovators.
- Users of marketing research information, such as product line managers, are demanding insights that are of more strategic value. Emotions research enables marketing researchers to deliver consumer insights that get an answer the strategic question, "So what"?
- The innovation manager sets strategy for the innovation team by identifying boundaries in which innovators can focus their expertise to innovate.
- Innovators are domain experts that require consumer and consumer product insights that both inspire their collective creativity and guide their decision making.
- The scientific method simply does not work well with innovation teams as an approach to learning. They require a more fluid, iterative approach built upon new sources of information from research designs that are more integrated and holistic, and that have dipped in to the new rivers of information flowing from the Internet.
- The challenge to integrate these new sources of information can be solved through the application of frameworks built upon the science of emotions.
- Story telling is an advancement in the delivery of insights that leads to greater relevancy to more members of the innovation team – facilitating more collaboration to achieve team goals.
- The innovation process can be visualized as two interconnected cycles: one cycle leads to learning for the purpose of developing an innovation strategy for brands, the other for the purpose of product development.
- Eight phases of innovation are included over these two cycles, with the knowledge from the preceding phase being the base of knowledge for the subsequent phase.
- Learning within each phase includes the process of divergent thinking (collective creativity) inspired by insights, followed by a process for convergent thinking with decisions guided by insights.

Chapter 4

The Science of Emotions

Emotion always has its roots in the unconscious and manifests itself in the body.

Irene Claremont de Castillejo

THE KEY TO DATA INTEGRATION

Chapter 3 discussed the multidisciplinary nature of innovation and how innovation teams work to meet the innovation challenges of the 21st century. The disciplines that innovation team members bring provide a base of professional knowledge that can be applied to develop innovation strategy and to implement that strategy to develop breakthrough products. A key factor contributing to the success of innovation teams is the capability to learn through the generation of consumer insights and consumer product insights. The nature of innovation requires a different approach to learning than the scientific method. It requires an approach that is more iterative and holistic, integrating a more diverse set of data into insights that can be woven into stories. This is where frameworks come in to play – they are the key to integrating into insights all these new sources of data collected through quantitative and qualitative hybrid research designs, or harvested from the new rivers of information flowing from the Internet.

However, as this chapter will show, not all frameworks are sufficient to meet the learning needs of the innovation team. Innovation teams require insights into how and why consumers behave, in order to be more successful in the development of breakthrough products. This chapter will first show the importance of emotions in consumer behavior – how emotions drive many consumer behaviors. Drawing from a number of scientific disciplines, a story will be told for what is presently known about the science of consumer behavior. The story will begin with what is known about our brain's structure and how this understanding provides some insight into human nature – why consumers often react irrationally and behave as herding creatures. It will conclude with the formation of a new thought process to integrate information generated through sensory and marketing research into insights. This new thought process will take the form of three different types of frameworks all built from what is known about the science of emotions that can be

Breakthrough Food Product Innovation Through Emotions Research. DOI: 10.1016/B978-0-12-387712-3.00004-9

applied to inspire and guide innovation teams to achieve breakthrough products.

FRAMEWORKS

Frameworks are not new to research and food product innovation. In the 1950s, the science of cognition was applied to develop a framework for product marketing. This framework was taught in marketing courses as marketing science. It was applied in the development of standard marketing best practices, such as how to increase awareness through a mix of communication channels to increase market penetration. It helped marketing researchers develop methods to gain insight into how consumers make rational decisions to purchase products. At the same time, the science of perception was applied to develop a framework for engineering and manufacturing food and other packaged goods. This framework resulted in the difference-testing and acceptance-measuring techniques used today by sensory researchers. It helped companies develop quality tolerance specifications and establish processes for better product development guidance based upon the judgments of trained panelists (e.g. flavor and texture profiling).

During the 1980s, the science of sensory pleasure was applied to develop a new framework for product optimization. Researchers applied this framework to identify what qualities of products drive pleasure and how to optimize the levels of these qualities to maximize sensory pleasure. This framework is the basis for many of the consumer research techniques and statistical modeling methods used today to design and develop pleasurable products.

These examples of frameworks have led marketing, sensory and consumer researchers to apply scientific disciplines in order to make sense out of complexity. They have led to new ways of generating information from consumers and also to new analyses of quantitative data. They form the basis for many of the standard practices used in product innovation and development. Innovation decisions are guided and creativity is inspired from insights based upon methods developed from these frameworks. Marketers approach innovation on the basis of cognitive science, seeking insights into how consumer behavior is driven by rational thought. Developers approach innovation on the basis of the science of perception and sensory pleasure.

However, consumers often do not act rationally. Sensory pleasure is often not the chief reason consumers try and repeat consuming or using products. In the stories of success and failure presented in Chapters 1 and 2, we saw how something has fundamentally changed in how and why consumers react to products. This calls into question the effectiveness of current innovation processes based upon existing research methods. These facts suggest that companies need to change their paradigm and approach to innovation. It also suggests the need for a new framework – one based upon not just the science of cognition, perception and pleasure, but also the science of behavior.

In this chapter, four interrelated frameworks will be defined, each a practical application to solve different parts of the data integration problem for innovation teams. The first will be an "appraisal framework" that will coalesce the science of emotions into a conceptualization for how to elicit the formation of different types of emotions. The second will be "information frameworks," definition of which will provide a conceptual structure to think about the types of information we collect through research in building knowledge about consumers. This leads into a discussion about "causal frameworks" and "dynamic frameworks" – i.e. new ways to structure the design of research and integrate information collected from research to generate insights for breakthrough innovation.

THE SCIENCE OF BEHAVIOR

Recent Scientific Advances

Sipping Cabernet Sauvignon provided Dr. Hilke Plassmann and her fellow researchers, at the California Institute of Technology, with rich data for studying the price of a product and its effect on the brain. Using MRI (Magnetic Resonance Imaging), researchers found that pricing information affected the pleasure centers of the brain. The subjects reported that the wines "identified as more expensive tasted better."[1] With MRI, researchers can see an image of the brain as people are exposed to different stimuli. This type of imaging provides insight into the different parts of the brain that control cognitions, perceptions, emotions, and the hedonics or pleasure areas of the brain.

Dr. Plassmann's subjects were given three different wines to taste with the price tags switched. For example, a $90 bottle of wine was presented with a fictitious $10 tag. The subjects were told the price points of the wine, but asked to focus on flavor and how much they enjoyed each sample. The research found that, in this test, the perception centers of the brain were unaffected by the price differentiation, but the emotion and pleasure centers were highly impacted – as well as their cognitions or interest in purchasing a certain wine.

Functional MRI (fMRI) was used in an earlier study to focus on the hippocampus part of the brain, where the memory centers and the emotions are connected. Subjects in this study were asked to sample Pepsi and Coke in a blind taste test. In some tests, they were told the brand (sometimes incorrectly) and in others they were given the tasting without the branding. What was found was that the brand-knowledge in the case of Coke had a dramatic influence on brain response and on behavioral preferences.[2]

1. L. Trei, Does a Wine's Pricetag Affect Its Taste? Stanford Graduate School of Business News, January 2008

2. S.M. McClure, J. Li, D. Tomlin, K. Cypert, L.M. Montague P.R. Montague, Neural Correlates of Behavioral Preference for Culturally Familiar Drinks, Neuron (44) (2004) 379–387

In the absence of brand information, subjects split equally in their preference for Coke and Pepsi. However, given several cups of Coke with some labeled Pepsi, the subjects showed preference for the cups labeled Coke.

These studies become relevant to the problem of achieving better product innovation, since valid insight can only come from a holistic view of the consumer experience. A holistic view includes emotions, perceptions, and cognitions and shows how they all play together. Without all three, you cannot truly understand the "whys" of consumer behavior. For example, if you use a blind taste test to isolate an essential element like perception, you are missing other key areas – you are not being holistic enough to generate valid consumer product insights.

From these research studies, we learn that sensory perception is only one of many factors contributing to hedonic (pleasure) and emotive response. Conversely, we learn that perception is only affected by the sensory stimulation of a product (e.g. the wine). In other words, perception occurs whether or not the subject was aware of it. Perception was not impacted by rational thought about the price-value of the wine. On the other hand, non-sensory information, that is, external information such as branding and pricing, greatly impacts not only our experience of sensory pleasure, but our experienced emotions. We also can see that cognitions about preference and behavioral motivations are influenced by both our experience of pleasure and the resulting formation of emotions, as well as by emotions formed from our memories.

In Martin Lindstrom's book *Buy-ology: Truth and Lies About Why We Buy,*[3] he describes research using fMRI to measure the effects of cigarette packaging warnings of smoking behavior. While subjects would rationalize that they were smoking less due to the warnings on cigarette labels, the center of the brain that is associated with desire (i.e. nucleus accumbens) was actually stimulated by the warning labels. This insight is surprising – it implies that cigarette warnings lead to *greater desire* to smoke among smokers.

In his book *Habit: The 95% of Behavior Marketers Ignore,*[4] Neil Martin describes in great detail the different centers of the human brain and how these relate to our behaviors. The gist of the book is a claim that 95% of our behaviors are elicited through learning that occurs unconsciously, rather than consciously – as we might like to think. The unconscious part of our mind is associated with brain activity from what many call the hindbrain – the brain's "basement," containing the pons, medulla oblongata, and cerebellum – and the limbic system – the brain's middle section which includes the amygdala (our emotional center) and the hippocampus and basal ganglia (the brain's central switchboard). Conscious, rational thought occurs through activity in our frontal brain – the cerebrum.

3. M. Lindstrom, Buy-ology: Truth and Lies About Why We Buy. Random House, Inc., New York, NY, 2008, pp. 240

4. N. Martin, Habit: The 95% of Behavior Marketers Ignore. Pearson Education, Upper Saddle River, NJ, 2008, pp. 191

Martin notes that living creatures have had a hindbrain for at least 200 million years, as evidenced by fossilized dinosaur skulls. We and other animals use hindbrains for very basic, repetitive unconscious behaviors such as walking. The limbic system is where memories are formed, habits are learned, and emotions are elicited. The hippocampus is responsible for creating, storing and retrieving memories. Our emotion centers shape what memories we choose to store and what we recall. The stronger the emotion, the greater the likelihood the event will be placed into memory. Memory retrieval is greater when events occur eliciting similar emotions, or with cues associated with those events. Basal ganglia are involved in learning habitual behaviors. When we repeat a behavior, even when it is a complex series of tasks, those tasks are encoded in our basal ganglia. Cues – or sensory perception triggers – associated with these habitual behaviors are also encoded, waiting to be activated upon cue.

These recent understandings of the brain – mostly derived through fMRI research – are shedding new light into the underlying drivers of consumer behavior. As has been described in experiments with subjects sipping wine, soft drinks, or observing cigarette packaging warning labels, the unconscious minds of ordinary consumers are constantly active. The emotions we feel during consumer experiences are glimpses into the continuous processes being carried out by our unconscious minds, sifting through vast quantities of sensory stimuli to sort out what events are worthy to be stored for future retrieval. Our emotions are clues to what habits are being formed.

While science is now suggesting that much of our unconscious learning is mitigated through the emotions and cues we associate with experiences, there is also evidence that conscious learning plays a role in consumer behavior. Evidence is strong that cognitive information, such as price points, does impact expectations. It is through these expectations that we are made ready to experience feelings in response to product experiences. These scientific advances suggest that our consumer behaviors are the result of complex processes that occur within both our conscious and unconscious minds.

The importance of these scientific developments should not be left for just scientists to ponder. These advances have huge practical significance to domain experts involved in innovation. What stands between the science of behavior and its practical application for innovation teams is the development of behavioral frameworks that enable the theories of behavioral science to coalesce into a useful tool for use by innovation managers, creative designers and developers, sensory scientists, marketers and marketing researchers.

THE PSYCHOLOGY AND THEORY OF EMOTIONS

Emotions as defined and utilized by scientists are complex and, in general, are not understood by innovators, marketers, developers, and brand owners. To get to the meaning of emotions, we start with identifying the emotive phenomena we can observe or measure. These are the occurrence and/or internalization of feelings associated with three areas: physiological changes, evaluative changes, and behavior "action readiness" changes.

Physiological changes include skin variations, heartbeat rhythms and other measurable phenomena. Evaluative changes are those reactions that we can see or hear, such as expressions or guttural reactions and sounds. The third area of emotive phenomena, behavior action readiness, was defined by Nico Frijda in his landmark book *The Emotions*.[5] Behavior action readiness refers to attention arousal – the reaction that occurs when something grabs your attention. Within this are "behavior action tendencies," which are emotions that prepare us to act. Behavior action readiness can be thought of as enjoyments and desires – things we seek that can cause us to move toward or away.

Emotions have discrete qualities that are easily distinguishable. For example, enjoyment is very different from pride. Emotions can be positive or negative. Pleasant surprise is very different from unpleasant surprise. Emotions have a time intensity, with an onset, a duration, and an end point. Emotions are projected onto a self, another person, a product, a brand, or an experience. You may project an emotion onto yourself: I feel sad or proud; or project it onto another person: I am angry at that person or that company. Or, an emotion could be projected at a product: I love this product.

Emotions are not like moods, which are long term states of disposition that cannot be projected onto anything. A mood of sadness is simply a persistent state of feeling that is not projected onto a specific event or object. A mood or temperament (feeling high or low) might last for days or longer.

Emotions are also different from attitudes, which are enduring states of disposition about products and other objects that are long-lasting. Attitudes toward products, brands, people and experiences can be enduring and difficult to change.

Emotion Theory

Although there have been many theories about emotions, which have been evolving for more than 30 years, the PAD (Pleasure, Arousal, Dominance) model, developed in 1974, is still widely used. The model does not look at emotions as discrete, but rather as the collective positive or negative direction of feelings.

More contemporary theories of emotions have been developed in the past 20 years. Many of these theories consider emotions to be a mix of discrete qualities of feeling. The ideas presented by Frijda, in his 1987 book, form the center point for many of these more contemporary theories. Recently, there has been considerable work published on *anticipated*

5. Nico Frijda, The Emotions. Cambridge University Press, Cambridge, MA, 1987, pp. 544

emotions, as distinctly different from *anticipation* emotions. The former are not true emotions, but simply expectations: I expect that I will feel comfort if I use this product. The latter are true emotions, such as the feeling of intrigue about anticipated experiences, or the desire to own or possess a product.

The generation of consumer and consumer product insights that get to the "whys" of behavior requires the adoption of the more contemporary theories that consumers are motivated by a mix of discrete emotions. The characterization of emotions as discrete provides a basis for understanding the different "eliciting conditions" that underlie the formation of emotions, and a framework to turn the level of different emotions up or down through innovation.

Topology of Product Emotions: Discrete Emotions and Eliciting Conditions

Various topologies have been developed for defining and characterizing feelings as discrete emotions. Emotions may arise in anticipation of an experience, e.g. feeling intrigue, or in response to an experience itself. Emotions may occur concurrently or in sequence, e.g. an initial feeling of enjoyment (taking pleasure from an experience) may shift to a feeling of satisfaction (the experienced pleasure was as expected). For example, you may feel anger when you discover that a brand has made a decision to change or eliminate your favorite feature or flavor. It may evolve into sadness when you reflect on the situation and realise your favorite feature or flavor is no longer available.

Behavioral psychologists have proposed a number of different emotive topologies with differing numbers of discrete emotions. In reviewing and applying different topologies, one topology in particular seems to provide sufficient granularity, not only to name a wide range of emotions, but more importantly to characterize a wide enough range of eliciting conditions to be of practical application for consumer product innovation. Eliciting conditions characterize emotions on the basis of how concerns and expectations are, or are not met, and how a specific emotion is projected onto an object (a product, brand, company, experience, another person or one's self).

A topology that specifically focuses on product emotions was developed through graduate research by Pieter Desmet, a behavioral psychologist involved in product design in The Netherlands.[6] This topology (Table 4.1) includes eliciting conditions that are specific to consumer products formed when consumers anticipate a product experience or have a product experience. It provides a definition of 22 basic emotions.

---

6. P.M.A. Desmet, Basic Set of Emotions. A Typology of Fragrance Emotions, in Fragrance Research 2005: Unlocking the Sensory Experience. Amsterdam, ESOMAR, 2005, pp. 134–145 (ISBN 928311780)

TABLE 4.1 Twenty-Two Product Emotions and Eliciting Conditions

Emotion	Eliciting Condition
Shame	Disapproving of one's own blameworthy action
Jealousy	Wanting what someone else has
Fear	Facing an immediate, concrete, physical danger
Anger	A demeaning offense against me and mine
Sadness	Having experienced an irrevocable loss
Pride	Approving of one's own praiseworthy action
Hope	Fearing the worst but yearning for better
Relief	A better goal realization (or concern match) than expected
Boredom	An unwanted lack of stimulation
Contempt	Disapproving of someone else's blameworthy action
Admiration	Approving of someone else's praiseworthy action
Disgust	Taking in being too close to an indigestible object or idea
Desire	An object calls for possession or usage
Disappointment	A lesser goal realization (or concern match) than hoped for
Love (liking)	Liking an appealing object
Dissatisfaction	A lesser goal realization (or concern match) than expected
Amusement	Some incongruity is solved
Stimulation	A promise for understanding through exploration or a new action
Satisfaction	An expected goal realization (or concern match)
Unpleasant surprise	An unexpected goal obstruction (or concern mismatch)
Enjoyment	Liking a desirable or pleasant event
Pleasant surprise	An unexpected goal realization (or concern match)

Based on Desmet, 2005[6]

This topology has proven to be extremely helpful in the design of holistic research, in strategic planning for product design and development, in the discovery of opportunities to improve existing or to develop new products, in the scoping of ideas into full product concepts, and in the establishment of innovation strategy for brands.

The Psychology of Emotions: Theory of Action Readiness (Frijda, Desmet)

This emotive topology fits well into the contemporary emotive theories that have evolved from Nico Frijda's ideas of "action readiness," as well as within the classical stimulus–response theory used by sensory scientists and perceptual psychologists to characterize perception and the formation of sensory pleasure. It expands upon perceptual psychology, by understanding what other factors contribute to the eliciting of emotions at the point of a product appraisal.

This theory centers on the moment of appraisal – the moment where sensory stimuli about a product (conceptually or tangibly) are captured and sent to the human brain. It is at this moment of appraisal where expectations and "cues" are formed in the mind of the consumer. These expectations are conscious notions for what might be the behavioral or emotional outcomes from the use of a product. Cues are the sensory stimuli at the very moment that lead to unconscious feelings about the product and outcomes from use. It is also at the point of appraisal where a consumer's concerns about a possible outcome come into play. This theory leads to a behavioral framework with inputs being the expectations and concerns of consumers, as well as the sensory stimuli that are used to not only form expectations that are contrasted against held concerns, but also lead to emotions from cues. The output or result of this appraisal is the elicitation of various discrete emotions characterized in the above topology.

In his paper, "A Multilayered Model of Product Emotions," Pieter Desmet[7] presents the schematic shown in Figure 4.1, which is helpful to visualize these inputs that lead to the formation of various product emotions at the point of an appraisal. This schematic was adopted for its simplicity and accuracy in characterizing the formation of emotions, as applied to understanding consumer product experiences.

Desmet expands on the definition of concerns. He defines "surface concerns" as the specific concerns or feature preferences desired at the point of an appraisal. These concerns are distinguished from "source concerns," which are defined in this chapter as the psychological goals, standards, attitudes, or ideals that consumers use to direct their general tendencies in taking action.

Product expectations differ from concerns. They are pre-experience notions about what the most likely outcomes are from use of the product being appraised. Expectations may be formed from within our unconscious mind, in the form of an intuitive "sense" for what is the most likely experience outcome from use of a product. We intuit what might be the experience outcome based

7. P.M.A. Desmet, A Multilayered Model of Product Emotions, The Design Journal 6 (2) (2003) 4–13

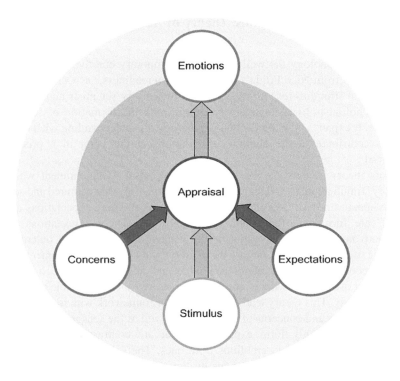

FIGURE 4.1   The basic appraisal framework as developed by Pieter Desmet.

upon the sensory stimuli we perceive at the point of an appraisal. Through the process of priming, we are programmed to automatically associate sensory stimuli perceived at the point of an appraisal with past memories and their respective emotions.

When we appraise a product conceptually, our feelings toward or away from that product are formed from emotions. Those emotions are in response to sensory cues, e.g. the visual or auditory cues associated with the product concept or picked up from the situational context that are associated with past experiences that were of sufficient impact to be registered (i.e. primed) into our unconscious memories. Also contributing to the formation of emotions is the level of expectations and concerns held at the moment of appraisal. Expectations and concerns lead to the discreteness and intensity of emotions, i.e. whether we will feel fear, hope, intrigue, desire or even disgust. If the concept provides content that leads to expectations not in line with past experiences (i.e. contrast with our expectations), we may feel intrigue in anticipating the experience. If our concerns at the moment are heightened, we may feel greater desire or disgust toward the product.

When we appraise a product experientially, these same expectations come into play and are contrasted with our concerns against our sensory perceptions throughout the product experience. Intrigue may lead to amusement. We may feel enjoyment or satisfaction. We may feel love or liking for the product. We may feel shame for our blameworthy behaviors or regret in not using an alternative after using a product. When an experience is not as expected, we may feel a pleasant or even an unpleasant surprise.

This behavioral framework provides those involved in innovation with the knowledge as to how emotions are formed during everyday experiences with products. Applying this simple behavioral framework creates new thinking for how to innovate. The framework provides us with new knowledge for how to more effectively communicate. By heightening "expectations" in line with consumer "concerns" through communications from a trusted source ("stimuli"), the consumer is made ready to act. This simple framework suggests that emotions, such as desire, can be heightened during a point of purchase appraisal by appealing to the surface concerns of the consumer, or by heightening expectations about the emotional outcome from use of the product. It provides a strategic framework for more effective messaging in advertisement or for package design. Likewise, anticipatory emotions, such as intrigue, can be heightened by creating an expectation that some incongruity (e.g. two seemingly competing concerns such as healthy and tastes good) will be uniquely solved through use or consumption of the product.

These emotions make us action-ready to use and/or purchase products. They drive our behaviors by eliciting emotions triggered by the unconscious process of storage of memories from past, associated experiences, and/or begin the formation of a new habit to take future action. Simple behavioral frameworks such as this open up the process of innovation, inspiring managers, creatives, and insight generators to consider new possibilities – all focused on a common goal – for the motivation of consumers to take action that are of commercial importance to the brand owner.

These resulting emotions often lead to conscious rationalization – where we become aware of our feelings and attempt to understand their cause. It is this process of rationalizing where most quantitative research begins to falter. Consumers may have difficulty in expressing their feelings in response to structured questions. Recollection of past experiences may not concur with reality.

APPRAISAL FRAMEWORKS

The Four Dimensions of Experience

The application of the theory of "action readiness" as a simple framework with concerns, expectations and sensory stimuli as inputs, and emotions as the output, is in itself too simplistic to be of practical significance in consumer product innovation. It can be considered akin to Albert Einstein's famous

TABLE 4.2 The Four Dimensions of Consumer Product Experiences as Noted from Three Different Disciplines (Behavior Psychology, Design, and Business Strategy)

Discipline of thought	Functional (Realm of effects, results)	Social (Realm of action, interaction)	Sensory (Physical realm)	Psychological (Psychological realm)
Behavioral Psychology (Pieter Desmet)	**Instrumental** • Relative to goals—facilitation or observation • Also consists of responses elicited by anticipation of ownership/usage • Satisfaction, dissatisfaction, desire, disappointment	**Social** • Relative to standards—approval or disapproval • Admiration, indignation, contempt	**Aesthetic** • Relative to innate and acquired preferences (attitudes) – liking or disliking • Attraction, disgust	**Surprise** • Relative to expectations and concerns – expected or unexpected • Pleasant surprise, unpleasant surprise **Interest** • Relative to challenge combined with promise • Boredom, fascination
Design (Patrick Jordan)	**Ideo –** • Values aspirations, style • Desirable outcomes	**Socio –** • Relations with others • Relations with society (identity) • Social status of ("self-social") identity	**Physio –** • Sensory • Aesthetics	**Psycho –** • Pleasure from mental, emotional reactions • Stimulation, engagement • Psychological state • Includes emotion/mood
Business Strategy (Clayton Christensen)	**Functional** • Related to utilitarian aspects of the product and what it does	**Social** • Related to the social aspects of product ownership and usage		**Emotional** • Related to the emotional aspects of product ownership and usage

equation $E = mc^2$. This simple equation was used as a basic framework upon which to capture the essence of the complex theory of General Relativity. However, this equation is much too simple to be of practical significance to ensure, for example, that a satellite is able to communicate or a rocket ship is able to blast off into outer space. The simple behavioral framework, with concerns, expectations and sensory stimuli as inputs, while fundamentally solid on a psychology basis, needs to be extended to characterize why consumers experience 22 possible emotions through the use of products. Like Einstein's simple equation, it needs a more robust theory to be of practical significance.

In fact, a common solution to extend this basic framework into a more robust behavior framework has emerged from three different disciplines: behavioral psychology, industrial design, and business strategy. Patrick Jordan[8] proposed a design framework including four basic considerations for effective emotional design. Jordan's work influenced Pieter Desmet,[9] a behavioral psychologist, to expand the theory of action readiness to consider these four same dimensions of human experience. It is also notable that Clayton Christensen, a Harvard business professor well known for his ideas on innovation strategy, agreed that innovation strategy should also focus on these same dimensions of human experience. Christiansen and Raynor, in their book *The Innovator's Solution*,[10] identified three of these same four dimensions as "jobs-to-be-done" that motivate consumers to "hire" products or to build relationships with brands. Their perspective is that innovation teams can use dimensions to develop innovation strategy to evolve brands.

These three perspectives can be unified as shown in Table 4.2. The four columns at the top are, in essence, identical to the four pleasures as described by Jordan, the four types of appraisal as described by Desmet, and the three brand purposes and product jobs-to-be-done as described by Christensen and Raynor. In the application of strategic planning, these last authors lump sensory as a functional purpose.

These four dimensions of product experience provide a basis for extending Desmet's simple model on how emotions are formed into a broader appraisal framework for understanding the "whys" for a wide range of consumer motivations. It will also serve to unify the diversity of thought inherent in innovation teams into an even broader framework for innovation operations. In essence, the appraisal framework about to be discussed sets the basis for making operational the science of emotions for innovation teams.

---

8. P. Jordan, Designing Pleasurable Products. Taylor & Francis, Philadelphia, PA, 2000, pp. 224

9. Desmet, Multilayered Model of Product Emotions

10. Clayton M. Christensen and Michael E. Raynor, The Innovator's Solution. Harvard Business School Press, Boston, MA, pp. 2003, 301

The Emotions Insight Wheel

This broader appraisal framework characterizes in more detail each of the four dimensions of appraisal. The simpler framework of appraisal is extended by considering the respective different types of inputs (concerns, expectations, and sensory stimuli) and outputs (appraisal-specific discrete emotions) specific to functional, sensory, social, and psychological appraisal. In addition, this framework can be extended by considering the impact of context on appraisals. This extension makes a distinction between more general "source" concerns and specific "surface" concerns that are expressed within specific contexts. The extended framework helps explain how expectations from past experiences and sensory cues perceived in specific contexts form the basis for emotions. The incorporation of information about context, surface concerns, expectations and sensory cues is sufficient to explain what emotions are at play, in a given situational context, for a specific consumer with given source concerns. It leads to the ability to build effective models that have a statistical or heuristic basis to predict or infer respectively which of the 22 possible product emotions are at play. Capturing this knowledge into a usable appraisal framework creates a powerful tool for strategic planning, creative design and development and behavioral insight generation.

The schematic in Figure 4.2 provides a visual representation of this more general appraisal framework. Each quadrant characterizes the sensory stimuli (cues), concerns, and expectations that are associated with the four dimensions of appraisal.

The outer ring (hope, fear, and intrigue) and center (desire and disgust) represent possible anticipation emotions at play. The second ring in from the outer ring represents possible experiential emotions at play for each dimension. The innermost ring characterizes the associated projection of the emotion, with the respective experiential emotion just outside the projection ring, or the adjacent anticipation emotion at the outer ring (if present). For example, moving from the center outward, the Emotions Insights Wheel can be applied to develop a pathway (or trace) to get from desire ("an object calls for possession or usage") to the underlying cues, concerns or expectations. For example, desire for repeat (i.e. action readiness) can be elicited through a functional appraisal where the experience outcome (e.g. "package opened easily") met the expectation (e.g. "packaging for this brand will open easily") for an important concern (e.g. "the package will open easily"). This framework also leads to the prediction that satisfaction will be low (and desire low) if this concern is not important or if this is not the expectation. If one had the anticipation emotion of "fear" that the package could not easily be opened and outcome was "package opened easily," then this framework would predict that the experience would lead to relief and heightened desire in anticipating using the product again.

In this way, this appraisal framework leads to a range of different pathways to achieve increased desire or decreased disgust. By putting the focus on

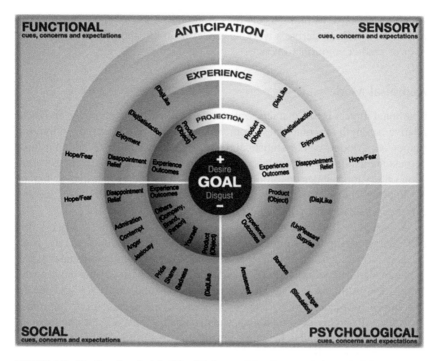

FIGURE 4.2   The Emotions Insight Wheel™ characterizing the expanded appraisal framework of anticipation and experience emotions associated with the four dimensions of product experience. (Please refer to color plate section)

emotions, this framework becomes a strategic tool to consider new ways for increasing a product's competitive strength. It extends the playing field in a new way. For example, most food products are developed conceptually to compete on a functional and/or sensory appraisal basis. However, it is known that many food products are tried and habitually used for how they deliver social-identity, leading to emotions such as pride and admiration. In other cases, desire (action readiness) to try can be elicited through heightened stimulation (i.e. intrigue) through messaging, or from an experience where the product elicits amusement or a pleasant (unexpected) surprise. As will be discussed in more detail in Chapter 5, these social and psychological types of appraisal lead to greater competitiveness for products and the brands they represent.

APPLYING APPRAISAL FRAMEWORKS

In order to apply this appraisal framework for innovation, one needs to derive an appraisal model that fits the consumer and context of the appraisal. This includes defining the underlying drivers of behavior for the consumer and the context within which the appraisal occurs. The consumer and context impact which cues, concerns and expectations come into play as model inputs. This makes it possible to derive a working model from the appraisal framework to infer or predict emotional impact.

Appraisal Models

Emotional impact is the characterization of specific discrete emotions and their relative strength as outcomes from a model derived from the appraisal framework. Models can be statistical when quantitative measures of inputs and outputs are available. Or, models can be heuristic, leading to inferences based upon observation under conditions where quantities and/or qualities as inputs are known. Both statistical and heuristic models can be applied to generate extremely valuable insights for innovation teams.

A particular model can be derived from the appraisal framework by first characterizing the source concerns at play. Source concerns are your behavior drivers that underlie actual behaviors. They include consumer goals (e.g. intended behaviors), beliefs (i.e. what consumers believe they ought to do or not do), attitudes (i.e. affect that is persistent and directed), and ideals (e.g. hoped-for outcomes). Source concerns are highly dependent on life roles, life values, life styles, cultural norms, and personality traits (e.g. promotional or preventative). Goals and beliefs tend to ladder up to surface concerns, attitudes and ideal to expectations.

Which source concerns lead to surface concerns and expectations depends on the context of the appraisal. This includes the environmental factors (e.g. ambience) and psychological factors (i.e. degree of depletion of energy–time-resources, mood, importance of task, alternatives) that impact surface concerns, and expectations – as well as cues that will be sensed from product and environmental stimuli. Also playing a role in enhancing concerns and expectations is the particular job-to-be-done by product in the given context of use. If the job-to-be-done by a milkshake is to provide enjoyment during a long commute to work, then one might seek a milkshake that melts slowly and delivers random bursts of extreme flavor to consistently break up the monotony of the commute. If the job-to-be-done is to provide comfort after a long day on the job, then one might seek a traditional flavor that takes you back (mentally) to a time when you were relaxed.

These feature preferences are surface concerns turned on by the context of the appraisal moment. They define the particular instance of an appraisal model, based upon the appraisal framework. As a consumer experiences a product, these different inputs can lead to a vastly different emotional impact, ranging from heightened enjoyment, to deep disappointment leading, to radically different desires (action readiness).

Figure 4.3 offers a schematic for an appraisal model. In addition to the consumer qualities (target consumer) and context, this model separates the sensory stimuli into two types: brand messaging (contributing to expectations) and intrinsic product qualities (contributing to product perceptions). This

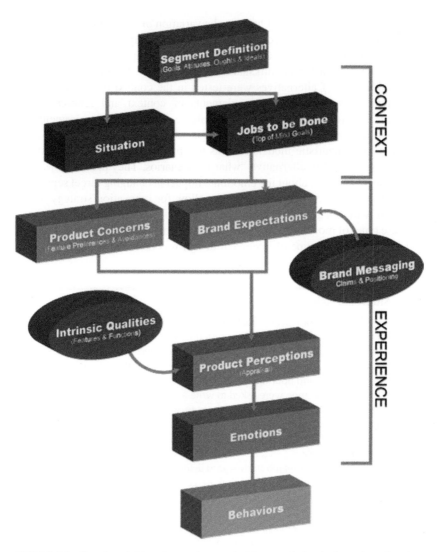

FIGURE 4.3 Causal model based upon the appraisal framework for the formation of product emotions motivating consumer behaviors.

separation enables building on appraisal models with heightened expectations from marketing communications. In the case of digital marketing, expectations might be impacted by peer-to-peer sharing, or messaging generated by a consumer seeking a product that can get a particular job done.

This model also provides a way to show the optimal sensory qualities that might be formulated to elicit the greatest emotional impact. Consider once more the milkshake example. A model can be generated from the appraisal framework by recruiting participants that "seek a milkshake to break up the commute monotony." This will help fix the variation of concerns in the model. Further, by providing branded messaging to all participants for the product concept the expectations can also be fixed as model inputs. This enables the holistic testing of different milkshake formulations to help predict or infer emotional impact. If alternatives were tested in the context of the car during a commute, then other unknown elements of context and random other consumer qualities would provide a holistic basis for measurement of emotional impact leading to emotions insight.

Statistical or heuristic models such as this provide a basis for data exploration, playing through alternative "what-if" scenarios. They provide a way to think through and learn about the emotional impact in changing target segments groups. These sorts of causal frameworks allow for innovation strategy to be developed with thought experiments to show "if-then-else" relationships.

These types of models are applied to integrate data into insights from holistic product designs. This type of integrated design was defined in Chapter 3. The following case study offers an example.

Case Study: Specialty Coffees

For the development of research methodology, InsightsNow conducted a study on specialty coffees. One of the coffees tested was a General Foods cappuccino product that was tasted by a variety of consumers who came into a central location as research participants. The participants prepared the instant coffee by opening the package and pouring it into a mug, and then adding hot water. They tasted the product and then were asked to write a story about their experience in preparing the product. We also asked them to write into the story how they envision they might (if at all) use this product. Each of the subjects was also engaged in a one-on-one interview to delve into more details about their experience and written story.

The purpose of this research study was to determine if a heuristic model might be formed from qualitative information to generate product insights. As an example, consider the written verbatim responses from one of the participants — "Nora."

"I was surprised that the product actually frothed when I added hot water. It was amusing to watch the froth grow over time. I didn't expect this from an instant cappuccino. I host 'coffee groups' in my neighborhood and would be proud to serve something like this to my girlfriends. I think they would enjoy it as much as I did."

In dissecting her response, different emotions can be inferred both from the adjectives she used to express her feelings, but also from the respective eliciting conditions or recognized cues she associates with each emotion within her story.

"I was **surprised** that the product actually frothed when I added the hot water."

"It was **amusing** to watch the froth grow over time. I didn't expect this from an instant cappuccino."

"I host 'coffee groups' in my neighborhood and would be **proud** to serve something like this to my girlfriends."

"I think they would **enjoy** it as much as I did."

These inferred emotions and eliciting conditions can be summarized as in Table 4.3.

In speaking with Nora, during the one-on-one interview, we were able to apply the appraisal framework to guide the interview to provide missing model inputs. We learned that she typically invites some of her girlfriends over and typically finds herself constrained by limited time to prepare. In addition, we were able to identify that she would love to offer a cappuccino drink to her girlfriends, but has no equipment to make an authentic drink. She is also seeking acceptance from her friends. She felt that the taste experience of the cappuccino needs to be acceptable (i.e. "To just taste as a cappuccino is expected to taste"). Further, she enjoys stimulating (novel) experiences and found the amusement of the frothing particularly stimulating.

This added information lead to a heuristic model that predicted the following:

Package appraisal: The picture on the package did not lead to heightened expectations that the product would be an authentic frothy cappuccino. Evidence for this was Nora's surprise at the foaming action and amusement in watching the frothiness grow after adding hot water during the subsequent preparation appraisal. The packaging did not offer any differentiating emotional impact such as intrigue or desire.

Preparation appraisal: Preparation led to psychological impact eliciting amusement and intrigue from the frothiness. While not explicitly stated, Nora felt functional satisfaction, as evidence that she has no cappuccino preparation equipment.

Consumption appraisal: The intrigue and amusement felt during the preparation experience most likely heightened the expectations for an authentic sensory cappuccino experience. As a consequence, she felt sensory satisfaction and sensory enjoyment as evident from her belief that in this context the taste experience needed to be just what is expected, and from her explicit statement about enjoying the product.

Conceptual appraisal of her envisioned use case: Her appraisal of the anticipated use case in serving her girlfriends suggests a high degree of desire (action readiness) to purchase for this use case. The brand and picture would cue the memory

---

TABLE 4.3 Nora's Inferred Emotions, Sensations and Pleasures Categorized within the Four Dimensions of Her Product Experience

Outcomes	Functional (*desirability*)	Social (*legitimacy*)	Sensory (*appeal*)	Psychological (*novelty*)
Emotions	Satisfaction	Pride	Enjoyment	Pleasant surprise Amusement
Sensations and pleasures	Convenience and authentic "cappuccino"	Anticipating: "coffee group" acceptance	Pleasurable sensations	Her observing frothiness growth

TABLE 4.4 Inferred Behaviors Motivated by the Four Dimensions of Nora's Product Experience

Action readiness	Functional (*desirability*)	Social (*legitimacy*)	Sensory (*appeal*)	Psychological (*novelty*)
Shopping behavior	x	x	x	
Preparation behavior	x	x		x
Consumption behavior		x	x	
Social behavior		x		x

of amusement, sensory enjoyment, and both sensory and functional satisfaction. Her anticipation of a positive social experience would lead to hope for acceptance by her girlfriends and would build an expectation that she would feel pride when serving her friends. This would heighten the expectations for the social appraisal, increasing the chance she will actually feel pride provided the social-related cues she gets from her girlfriends elicit positive emotions.

This heuristic model resulted in the generation of the following inferences about the behavior drivers that lead to action readiness for Nora (see Table 4.4).

This case study demonstrates how this causal framework can be applied to generate insights from research as simple as listening to a one-on-one interview. While the captured data is incomplete, you can apply the framework to infer missing aspects of the participant's experience. Imagine what can happen if you start filling in this grid for a large number of research participants.

With behavior-driven innovation, we can go one step beyond this to build a way in which we can see the emotional reach that products can deliver to markets. We can do this by understanding the individual's emotional connectedness to a product, along with their action-readiness in the encounter of the product. This allows us to identify utilitarian, social, sensory, and psychological-driven emotions with which we can build a profile of the product's impact on the individual. Furthermore, by rolling this emotional fingerprint up to a whole market, you realize that some people may be more about novelty; some may be more socially connected; some may find individual sensory appeal; and some may see more utilitarian value jobs to be fulfilled.

The bottom line is that the more emotional impact a product can deliver, spread across all the elements of the grid shown above, the greater emotional reach this product will have – with a result of greater success in the market.

Dynamic Models

This leads us to the final application of appraisal frameworks, the building and use of dynamic appraisal models.

Over a consumer's lifetime of experiences, they will invariably encounter the same product over and over again and thus build a cascade of experiences. In each situation, prior appraisals may form new memories, possibly changing expectations and concerns. These expectations and concerns from prior appraisals will, therefore, have an effect on the emotional impact of the subsequent appraisal.

One type of dynamic appraisal model is used to generate insights into how emotional impact changes over the same type of repeat appraisal. For example, with Nora, if the novelty wears off quickly and the cappuccino product no longer fills the job-to-be-done with her coffee group, then the sensory appeal may also fade and you might show a very short time footprint for her. These types of models are used to integrate data into insights from data collected from holistic use case designs, as defined in Chapter 3.

Another type of dynamic appraisal model is used to generate insights into how impact changes as consumers iterate through different moments-of-truth appraisals with the same product. An application of this is to generate insights into how the conceptual appraisal of an advertisement bears on emotional impact during subsequent appraisals, such as product preparation and/or consumption. The output from the conceptual appraisal might be found to heighten desire that increases the expectations for the outcome of the product preparation and/or consumption appraisal. These types of models are applied to generate insights from data collected through moments-of-truth designs, as also defined in Chapter 3.

As consumers encounter and appraise products in different contexts, the behavioral tendencies and readiness can be predicted by building and applying these types of dynamic models. By understanding changing source concerns, surface concerns, and product expectations over a sequence of experiences, these models can be applied to gain insights into behavioral dynamics. This may be applied to gain insight into use dynamics (e.g. learned or loss of liking), purchase dynamics (e.g. product switching), and social dynamics (e.g. short term fades and fashions).

Causal dynamics require you to think about why the consumer might be buying a product time after time – and then suddenly become bored with it and change their buying behavior. These sort of dynamics cause us to think about how emotional impact changes: how expectations change; or a product may change; or people's concerns may change. There may be change in the environment, or in a consumer's society that changes people's attitudes.

With constant change, you must ask how do you design products to not only have an emotional impact the first time, but to also have long-lasting sustainable impact? These and other questions will be addressed in the chapter – Breakthrough Innovation.

Key Points

- Recent understandings of how the brain works – mostly derived through fMRI research – are shedding new light on the underlying drivers of consumer behavior. These scientific advancements suggest that most consumer behaviors are the result of complex processes that occur within our unconscious minds. This scientific knowledge is distilled into a number of different conceptual frameworks of consumer behavior application in product innovation.
- The science of behavior links motivation for many consumer behaviors (i.e. eliciting conditions) to the eliciting of specific, discrete emotions defined as emotive topologies. A topology proposed by Pieter Desmet characterizes 22 discrete emotions and their respective eliciting conditions.
- Emotions theory distinguishes between anticipation and anticipating emotions. The former are distinctly different from experiential emotions, in that they are formed through the process of considering, rather than actually using a product. These are different from anticipating emotions – which are not true emotions, but simply an expectation that one would feel a certain emotion in a given situation.
- Emotions theory claims that discrete emotions are elicited from the contrast of surface concerns and expectations of relative importance, within a given situational context from an experience outcome.
- An extension of emotions theory into an appraisal framework involves the consideration of different cues, concerns, expectations as inputs and discrete emotions as outputs that occur under four different types of appraisal: functional, sensory, social, and psychological.
- The appraisal framework unifies thought diversity from innovation managers (strategists), innovators (designers) and researchers (insights providers).
- The Emotions Insight Wheel is a visual representation of the appraisal framework to apply emotions research generating insights for the innovation team.
- Appraisal models that predict and infer emotional impact can be derived from the appraisal framework by considering the behavior drivers (i.e. source concerns) and elements of context (environmental and psychological factors, and jobs-to-be-done) at play that turn on surface concerns and expectations, and the brand messaging and intrinsic sensory properties of the product experience. These models help integrate data collected through holistic product designs into insights.
- Dynamic models predict and infer change in emotional impact. There are two types of dynamic models. The first leads to insights into change in emotional impact over repeat appraisals on the same type of product experience. This type of model helps integrate data collected through holistic usage case designs. The second leads to insights into change in emotional impact over different moments-of-truth appraisals. This type of model helps integrate data collected through moments-of-truth designs.

Chapter 5

Emotions Research

Knowledge is of no value unless you put it into practice.

Anton Chekhov

CUTTING THROUGH THE CLUTTER

The spark for innovation comes from knowing the desires of the consumer and applying that knowledge to inspire creativity, or to guide decision making. To cut through the clutter of the crowded retail grocery outlet, the innovation team must be able to know how to become differentiated, and focus on applying their professional know-how and new knowledge gained explicitly in the form of research information.

In Chapters 1 and 2, an argument has been made that a chief factor for product failure today is the inability of companies to know how to respond effectively to market change. The topic of this chapter is how to improve the speed of learning about the "whys" of consumer behavior through emotions research. Speed for learning is essential for two reasons. First, the dynamic marketplace today requires speed in sensing and responding to consumer change. The need for this speed in response was seen in the wake of the 2008 recession that resulted in rapid consumer change. The other reason is that speed in learning translates into research efficiency. Research efficiency is defined as the degree of learning per resource spent. As we will see in this chapter, faster learning can accelerate learning while reducing research cost. Further, greater research efficiency enables the whole innovation process to become more efficient – more focused – and more successful.

In Chapter 3 an appraisal framework was described to model the causes and effects that elicit various discrete emotions when consumers appraise products conceptually or experientially. Underlying these emotions are specific qualities that characterize the consumer (i.e. goals, beliefs, attitudes and ideals that underlie behavior intentions) and aspects of context that define the appraisal moment (jobs-to-be-done, environment ambiance, psychological context). These measurable aspects of the consumer and context form the basis for basic model inputs such as expectations, surface concerns, and sensory cues that characterize an experience that elicits felt and portrayed expressions of emotions.

This appraisal framework provides a common ground for the innovation team to work toward common goals. It gives marketers a basis for behavioral characterization of consumers to discover opportunities for brands. Designers have a basis to translate those opportunities into conceptual and tangible product specifications that cue emotions within a broad reach across target market segments. Developers have a basis to prototype, formulate, engineer, mock up, and deliver a final product that delivers against the defined opportunity.

This chapter begins the discussion of how to achieve breakthrough innovation through emotions research based upon this appraisal framework. This includes the thought processes and methods by which researchers help innovation managers (strategists) and innovators rapidly learn. This chapter also sets the tone for the rest of the book, where we will delve into specific methods and techniques for each phase of the behavior-driven innovation process.

CONSUMER BEHAVIORS

The information presented so far about the science of emotions is largely theoretical. In our roles as innovation strategists, researchers, and innovators, we need to move from theory to practice. This does not mean the conceptual ideas for this framework need not be understood. In fact, it is essential to understand the basic concepts behind the appraisal framework.

Behavior-driven innovation begins with considering the different types of consumer behavior that might impact the consumption of a food product. There are four basic types of behavior important to drive the success of food products. These are summarized in the "4 S's of Consumer Behavior" (Figure 5.1) These behaviors are the Sensing behaviors that underlie all consumer habits; Sharing behaviors associated with our nature as herding

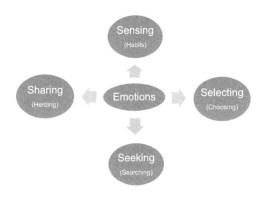

FIGURE 5.1 The 4 S's of Consumer Behavior: Sensing, Selecting, Seeking, and Sharing.

creatures; **S**eeking behaviors as prosumers (i.e. searching for brands that will work for us); and finally, **S**electing behaviors associated with choosing products to purchase and use. Each of these behaviors is motivated by the formation of emotions that make consumers "action ready" when the right cues and conditions present themselves. Emotions research is used to generate insights into how to motivate change in these respective behaviors through the lens of the appraisal framework.

Sensing Behavior

The most prevalent consumer behavior involves the habitual use of products. As earlier noted, some behavioral psychologists claim as much as 95% of all consumer behavior is habitual. Sensing behavior arises from sensory cues in the environment, the priming of new sensory cues at trial, and the reinforcement of sensory cues and associated habits through repetition. This behavior is called sensing behavior to make the point that habits are formed and reinforced by these sensory perceptions – most of which go unnoticed by our conscious mind, being driven by our unconscious mind. The appraisal framework can be applied to help innovation teams consider how to form and reinforce habitual use of products.

Innovation teams must first of all understand how habits are formed. This involves the process of priming, where high positive emotional impact at trial leads to the storage into memory of the emotions and sensory qualities associated with the experience.

For example, every parent can probably relate to the experience of their children first being introduced to ice cream. In tasting ice cream for the first time, children experience a sudden increase in pleasure, stemming from the sweetness, temperature, and texture. These sensory qualities, and the emotion of sensory enjoyment, are all stored in the brain as a lasting memory for later recall. The memory may also include a number of sensory stimuli, such as the spoken word "ice cream," the visual form of the ice cream, even the location of the event (i.e. a specific restaurant). All these environmental sensory stimuli associated with the experience, the qualities from the taste profile, and the memory of the experience itself are primed into the child's memory along with any associated emotion(s), e.g. sensory enjoyment.

These sensory qualities, memories, and emotions come back into play whenever the same sensory stimuli present themselves as cues. When the parent drives by the same restaurant, or the word "ice cream" is uttered, the reaction by the child is quick. The child's demand for a repeat ice cream experience is motivated because they feel once more the sensory enjoyment leading to desire.

This example of a child's experience is not too different from the classical psychological experiment generally known as "Pavlov's Dog". This classic experiment ("ring the bell and the subject becomes action ready") occurs every day in the life of not only children, but every consumer. The appraisal framework is essential to understanding how we build into products sensory cues (i.e. how to "ring the bell") to motivate consumers to become action ready to engage in sensing behaviors that drive product repeat.

Seeking Behavior

The second type of consumer behavior is to seek out – i.e. to search for alternatives. This involves rational thought. However, the triggering of seeking behavior requires the disruption of the unconscious mind – i.e. to break existing habits.

Disruption is critical to behavior change. It occurs when a consumer finds themself in a new situation, when an event changes their life, or when they are exposed to new information. These changes disrupt sensing behavior by changing consumer concerns and expectations such that environmental cues no longer trigger emotions that lead to habitual behaviors. The new situation, event or information forces the conscious mind to focus on appraising the situation through rational thought.

Disruptive behavioral change is most dramatic when it impacts a change in source concerns. This can result from a disruptive change in life role – e.g. the first time a new mother finds herself shopping for baby food. It may involve a change in beliefs from a disruptive event, e.g. the first time one is faced with a health-related concern requiring seeking out new food alternatives. It may also involve a shift in ideals through exposure to a new product, e.g. the "raising of the bar" in expected value or feature preference. Further, it may occur as a result of new information that impacts a consumer's attitude about a currently used brand, e.g. the awareness that a brand is not being authentic or truthful. In each of these cases, disruption occurs when a change in source concerns disrupts habits by eliciting new anticipation emotions. These anticipation emotions result in avoidance or hesitancy for use of an incumbent product.

Disruption leads to seeking behavior to find an alternative to fulfill concerns. It provides a behavioral basis for innovation teams to establish goals. If your goal is to take market share away from your competitor through innovation, you need a behavior-based strategy to disrupt the unconscious mind, i.e. to make the conscious mind more aware that it needs to pay attention to new information. This involves knowing how to switch the minds of prospective new consumers from "autopilot" to their conscious minds so that they can rationally consider choice alternatives.

Once disruption has occurred, seeking behaviors are played out in a wide range of different situational contexts, ranging from in-store shopping to at-home Internet searching. The form of seeking behavior is mostly a question of immediacy in access to trusted sources of information that a consumer seeks to influence their search. The Internet and the immediacy of information through smart phones have changed the nature of seeking behavior. Consumers may "text" trusted friends, post questions out on social networks, or seek out alternatives through the use of various search engines. The mobility of the smart phone means that consumers can access information anywhere, anytime. This change in how consumers engage in seeking behavior has dramatically changed the behavioral landscape for innovation marketing.

Seeking behavior is one of two key pillars for an emerging new model for market dispersion – i.e. how awareness and trial of new products is dispersed throughout a population of consumers. The other is sharing behavior, to be discussed below. Digital marketing is growing in importance as an essential aspect of the marketing mix for brands. It includes tactics such as introducing key words into Internet search engines that provide links to digital advertisement (e.g. YouTube video), or to informational content that is designed to disrupt or reinforce habits.

In 2006, Dorito's "Crash the Super Bowl" contest took the No. 1 spot on YouTube for most viewed Super Bowl commercial.[1] This digital marketing campaign inspired consumers to post videos on YouTube. This campaign heightened awareness for the brand, resulting in a spike in Doritos sales. In 2009, they upped the prize money to $1 million for the winning YouTube video and further improved their brand image and sales.[2]

Digital marketing is not only an "end game" consideration for innovation teams, but is an essential element in innovation strategy. Therefore, it should be considered as an element of strategy at the front end of the innovation process. The innovation team should pay attention not only to the cues used in building a digital marketing context to motivate trial, but also to how to motivate consumers to become engaged in seeking out information sources from a brand. An example of how innovation teams are using digital marketing is through the application of a technique called "crowdsourcing."

Consider the case study for Glaceau's Vitaminwater brand.[3] In this case, Vitaminwater developed an application (Flavorcreator) that could be downloaded off their Facebook fan page (see Figure 5.2). According to Jason Harty,[4] this application engaged "tens of thousands" of consumers for an average of 7 minutes. The application engaged consumers in a number of ways:

1. View relevant conversations around the Internet about Vitaminwater.
2. Vote on flavor combinations of interest.
3. Learn about the functional benefits of Vitaminwater and learn about themselves through an interactive game.

1. A. Lee, HP and Amazon Tap into Crowdsourcing for Ads, August 25, 2009, Fast Company.com. www.fastcompany.com/blog/anne-c-lee/green-room/hp-and-amazon-tap-crowdsourcing-ads

2. Press release: Doritos Registers Highest Brand Improvement Score in comScore's 2009 Super Bowl Survey, February 5, 2009, comScore, Inc. www.comscore.com/Press_Events/Press_Releases/2009/2/2009_Super_Bowl_Survey

3. C. Dillow, Crowdsourcing 10: Why Vitaminwater's Facebook App Can't Lose, September 10, 2009, www.fastcompany.com/blog/clay-dillow/culture-buffet/crowdsourcing-101-vitaminwaters-facebook-app-goes-beyond-fans-favori

4. J. Harty, Vitaminwater® connect(s) with our consumers. Presentation at ARF Industry Leader Forum – Putting Listening to Work, January 28, 2010, San Francisco

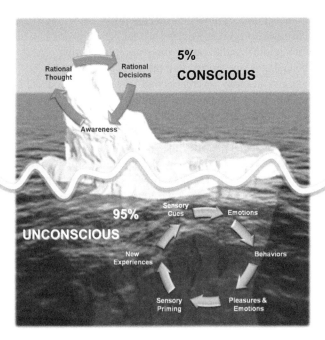

FIGURE 5.2 Interactions between the unconscious and conscious mind with regard to consumer behavior. (Please refer to color plate section)

4. Help create the ideal label.
5. Submit an ideal bottle and vote for it and others (created by other consumers).

This provided key innovation information to researchers resulting in the submission of 40,000 unique labels. Simultaneously, the loyal fan base was increased by 174%.

This "crowdsourcing" technique uses the prosumer's natural desire to seek out and support brands that will work for them. Crowdsourcing techniques engage large groups of consumers (defined as online communities) to interact through the Internet within a controlled context for the research purpose of creating information about their behaviors. Crowdsourcing techniques enable the capturing of information on not just sensing and selecting, but also seeking and sharing behaviors.

Joel Rubinson, former Chief Research Officer for the Advertising Research Foundation and current president of Rubinson Partners, summed up the opportunity for food companies with the following statement:

It's one thing to be excited about listening, yet quite another to apply listening in ways that impact brand performance, marketing and advertising. The willingness of brands and companies to organize around listening is critical to maximizing its potential.

Digital marketing techniques such as crowdsourcing provide for better "listening" into not just the performance of brands, marketing, and advertising, but also to generate insights into the behaviors of consumers and into the experiences that they seek.

Selecting Behavior

Selecting behavior can be habitual or rational. When it is habitual, the mechanisms discussed above with regard to sensing behavior apply to the context of choice. However, in the cases of disrupted habits, the rational behavior of seeking leads to rational, or a mix of rational and emotions-driven selecting behaviors. Rational thought (i.e. engagement of the conscious mind) occurs more frequently when the stakes are high for a decision and the individual feels accountable for making the decision. Rational decision making is also heightened when the complexity of the task is difficult, such as when there are many choice alternatives.

Consider the shopping experience for a new mother when first faced with the task of purchasing baby food. She will most likely have high source concerns that she should be a responsible mother. These source concerns might be apparent in understanding her other source concerns (goals, beliefs, attitudes, and ideals) in her new life role as a mother. In addition to feeling personally accountable for this important decision, she may find the task difficult in selecting from so many baby food choice alternatives. This would motivate seeking behavior.

The appraisal framework leads to new insights into the "whys" of selecting behavior. Based on her new role as a mother, it should not be concluded that her selecting behaviors are 100% rational. In fact, her past experiences leading up to this set of appraisal moments will most likely impact her choice decision. She may experience a flood of different feelings associated with perceived sensory cues on baby food package front labels. Some cues might elicit feelings of satisfaction or enjoyment from past associations from her own consumption experiences. Other cues in flavor combinations might elicit intrigue or even disgust. Her fears, hopes, intrigues, desires and even disgusts may impact her seeking behaviors and her ultimate choices.

She may feel emotions elicited from cues in the store environment, or cues on the front panels of packaging. These cues might have their origins in latent, unconscious priming that occurred through recently seen advertising, or be based on experiences from her childhood. They may even bring associated memories into her consciousness. Emotions play an important role in shaping and filtering her rational thoughts as she appraises different alternatives.

In this way, selecting behavior can involve a complex set of interactions between the unconscious and conscious mind. Her impressions (feelings from these cues while seeking) impact her rational thought process in much the same

FIGURE 5.3 The Flavorcreator application displayed on Facebook used in "crowdsourcing" methodology for the Vitaminwater brand. (Please refer to color plate section)

way as her rational thoughts focus her attention on new information and choice considerations that cue different emotions.

This process is portrayed in Figure 5.3, a schematic of how the unconscious and conscious minds can interact. The disruption needs to present new information that tells the unconscious mind that the concerns and expectations are no longer relevant such that what cued positive motivating emotions no longer applies to the appraisal moment. The game has been changed.

It is the seeking and selecting behavior – enabled through disruption – that leads to new opportunities for innovation teams to motivate choice for new alternative products. What is often missing in the thought process among innovation team members is a conceptual understanding of how to apply the emotions framework – to utilize the same information that disrupts the unconscious mind to create an argument for the conscious mind to consider selecting a new product among new choice alternatives.

What has changed in today's markets is that the flood of new information reaching consumers has increased the pace of disruption. This has led to the erosion of loyalty. Even when companies have succeeded in getting consumers to try a new product, they are not always successful at starting new habits that lock in an innovation team's product as the new incumbent. This "locking in" of a product, as the new incumbent, requires that the experience leads to strong enough positive emotion to re-prime sensory qualities for future use as cues that will start a new cycle of habitual behavior. This re-priming may lead to new meaning for the same sensory qualities that now cue a new set of emotions directed at the new incumbent and away from the old.

Sharing Behavior

The success of food products is not only dependent on the seeking behavior of prosumers, but also sharing behavior. In their book, *The Influentials*,[5] Jon Berry and Ed Keller back up the statement "when Americans make decisions today it is a conversation" with facts. According to Berry and Keller, word of mouth is important as the "best source" in 73% of consumers for "New meals, dishes to try." This can be compared to 24% for advertising (49 point difference). Malcolm Gladwell in *The Tipping Point*[6] tells the story of how market dispersion through conversation resulted in creating the Hush Puppies fad. It started with a few "mavens" (early adopters) who started wearing Hush Puppies. These few shared their experiences and influenced "connectors" (people who knew 200 or more people by first name) to start wearing them. These connectors influenced thousands, providing a tipping point for the fad to take off.

Internet social networking has further accelerated market dispersion such as this. The mass collaboration in peer-to-peer consumer sharing in social communities is not just an important phenomenon for marketers. It has implications for the whole innovation team.

In his book *Herd: How to Change Mass Behavior by Harnessing Our True Nature*,[7] Mark Earls offers perhaps the clearest perspective for how sharing behavior is having a significant impact on food product innovation. He views sharing as a consequence of our true nature, i.e. we engage in social behaviors "not just in a social context, but for social reasons." The following quote from his book is a good summary of this point:

We are community-minded and not selfish as certain political thinkers would have it; community-minded in this most important sense – we are a community species: we want to be together; we are made to be together; we are made by being together and we are made happier by being together.

His statement about people being happier when they are together suggests an emotional basis for social engagement.

5. E. Keller, J. Berry, The Influentials: One American in Ten Tells the Other Nine How to Vote, Where to Eat, and What to Buy. The Free Press, New York, 2003

6. M. Gladwell, The Tipping Point: How Little Things Can Make a Big Difference. Time Warner Book Group, New York, 2002

7. M. Earls, Herd: How to Change Mass Behavior by Harnessing our True Nature. John Wiley & Sons, Chichester, 2007

The appraisal framework, and the idea of action readiness, suggest that emotions are the underlying motivator for sharing behavior. This is true. However, there is something more fundamental about human nature that drives social behavior. As Earls pointed out, our true human nature is very apt to think about and conceptualize experiences in social contexts. Therefore, it is not unusual to hear research participants express their feelings through stories that have a social context.

Recall "Nora" from the specialty instant coffee case study in Chapter 4. She expressed a thought about sharing her coffee experience with her girlfriends. Her motivation to consider sharing her coffee experience had its roots in her initial trial (as a research participant). She felt functional-satisfaction from the convenience of preparation, sensory-enjoyment from the cappuccino taste, and surprise and amusement from the frothiness. However, Nora also expressed her desire to possess the coffee product through a thought that came quite naturally to her – a social context where she conceptually appraised the product. At that appraisal moment, she imagined this case and felt pride. This more than likely made her action ready to seek out the product, select it, and share it with her girlfriends.

Sharing behavior through online social networking not only drives influence, it also disperses news about ideas, viewpoints, and attitudes at amazing speed. For these reasons, Earls suggests we actually redefine the traditional definition of what is a "market." Traditionally, markets have been defined in terms of target population characteristics such as demographics and psychographics, or as quantified purchase volume, market value or market share. Earls defines markets in terms of who the market thinks they are; who the market wants others to think they are; who the target is influenced by; and the volume, value, and sharing of peer-to-peer interactions.

Sharing is perhaps one of the most important behaviors for innovation teams to consider using to drive new product success. By positioning products to have a social purpose and by building into products cues that elicit emotions from the social dimension of the Emotions Insight Wheel (see Chapter 4), the underlying motivation exists for the social side of human nature to drive new product success. However, as we saw with Nora, any form of positive emotion impact can provide the genesis for sharing behavior.

EMOTIONS RESEARCH INFORMATION

The common thread in driving sensing, selecting, seeking and sharing behavior is emotional impact. As these behaviors are key to innovation success, it is essential that innovation teams know how to conduct emotions research. Yet, as the appraisal framework proves, the factors involved in the formation of emotions are complex. This leads to a question of how to best apply the theory of the appraisal framework into practical application. An answer to this question starts with how you conceptualize and organize research information.

A conceptual thought about research information was introduced in Chapter 3 as the "Learning Cake." This schematic portrays how research information is integrated into insights to extend and apply learning. Further, the "Innovation and Development Learning Process" was introduced to conceptualize innovation team learning from one phase to the next. The knowledge from a prior phase in the process becomes the base for learning in a subsequent phase. Yet, neither of these schematics includes elements of the appraisal framework. What is needed is a way to conceptualize learning throughout the innovation and development process – through the lens of the appraisal framework.

Behavior Pyramid™

A natural way to conceptualize learning over any process is to build a pyramid. In fact, the research information required to apply the appraisal framework can be conceptualized as a pyramid with four levels. These four levels are, from bottom to top, consumer qualities, situation context for experiences, the actual experience, and resulting behavior (or action readiness). In order to understand the "whys" of consumers' behavior, it is important to first understand what fundamental characteristics or qualities of people are associated with their respective behaviors. This understanding narrows down the possibilities from the situations that consumers may find themselves in and what jobs they might seek being done by products. This leads to a more fundamental understanding of what are consumers' surface concerns and expectations. This leads to a better understanding for what they feel in response to stimuli in given contexts, and their associated behaviors or action readiness.

At the bottom of the Behavior Pyramid (see Figure 5.4) are qualities that characterize a consumer's identity. This ranges from traditional demographic information (i.e. population statistics such as gender, age, income, and education level), to measures about cultural and personality. The bottom level also includes information about a consumer's source concerns. This type of information can be readily collected through standard survey research.

A key piece of information being collected today from the river of information are source concerns associated with brands, i.e. brand attitudes – a new technique for gathering that information, which is gaining popularity today is sentiment analysis.[8] This technique samples peer-to-peer chatting from target online social communities and applies linguistic models (text analytics) to identify the specific language being used to characterize attitudes. Sentiment analysis can also characterize the affective state (type of attitude – or even emotion) and the direction of affect (positive or negative).

---

8. http://en.wikipedia.org/wiki/Sentiment_analysis

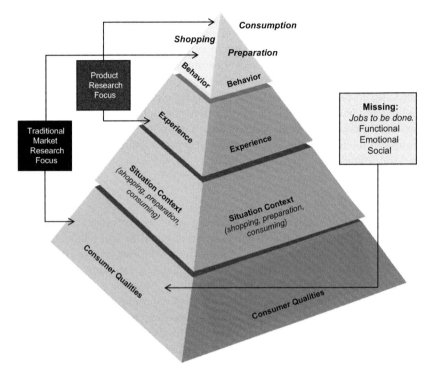

FIGURE 5.4 The Behavior Pyramid™ with the four levels of learning for consumer behavior (consumers, situational context, experience, and behavior). (Please refer to color plate section)

The second level up is situational context. This information characterizes situations within which consumers encounter and appraise products. As discussed earlier, these qualities include information about the environment (e.g. ambiance) and psychological states of mind (e.g. mood) of the consumer, including the jobs-to-be-done for a product (e.g. a milkshake helping make the commute more interesting) in that context. This type of information is collected through ethnography and diary, or journaling techniques that characterize context when consumers appraise products conceptually (e.g. shopping or watching TV) or experientially (e.g. preparing to use or consuming). This type of information is becoming more sophisticated, with the emergence of online "immersive research" techniques that task engaged consumers to collect pertinent information in pictures, and video clips that can easily be uploaded on a website.

The third level up is the consumer experience with the product. Experience information includes measures that characterize the stimuli of the product, surface concerns and expectations, and response measures by engaged consumers during and after product appraisal. Measures of stimuli can range from elements of a concept or advertisement, elements of a package, and sensory qualities of the food itself. This includes technical descriptive analysis methods,[9,10] which provide quantitative measures about food qualities that the consumer may or may not perceive, and which may or may not be sensory cues of which consumers are consciously aware or unaware.

Different techniques exist to measure surface concerns and expectations ranging from "check-all-that-apply" boxes on questionnaires, to stories told verbatim or through self-reported open-ended text in questionnaires. Expectations are generally collected in response to stimuli such as names and logos for brands or as concepts boards.

Sensory response measures of perception and pleasure involve classical consumer research methods. These techniques are covered in detail by Stone and Sidel[9] and Mailgaard et al.[10] They involve typical question-and-answer, stimulus–response, and quantitative research techniques.

Measures of cues and emotions require newer, less-known techniques. The fact that sensory cues and emotions are elicited from the unconscious mind compromises the use of classic sensory and marketing research methods that require rational thought. Therefore, these measures must be collected through indirect metrics. Visual cues in the form of words, pictures, and other imagery can be measured by tracking eye movement, or by tasking engaged consumers to recall, highlight, or select with their computer mouse visual stimuli that are of interest or of meaning to them. Indirect measures of emotions can include facial recognition by trained observers,[11] vocal recognition,[12] and cross-modal matching of feelings with visual and auditory "emoticons".[13]

Of emerging interest to researchers are the applications of biometric measures[14] and quantitative neuroscience metrics[15,16] to measure consumer response to stimuli. Biometrics includes physiological measures such as heart rate, skin temperature, and respiration rates that have been known to be associated with emotional responses. Neural measurement has provided researchers with a new (indirect) measure of change in blood flow to different parts of the

9. H. Stone, J. Sidel, Sensory Evaluation Practices, third ed., Elsevier, Amsterdam, 2004

10. M. Mailgaard, G.V. Civille and B.T. Carr, Sensory Evaluation Techniques, third ed., CRC Press, Boca Raton, FL, 1999

11. D. Hill, Body of Truth. John Wiley & Sons, Inc., New York, 2003

12. T. Johnson and K.R. Schere, Vocal Communication of Emotions, in M. Lewis, J.M. Haviland-Jones (Eds.), Handbook of Emotions, second ed., The Guilford Press, New York, 2001, pp. 220–235

13. P.M.A. Desmet, C.J. Overbeeke, J.J. Jacobs, When a Car Makes You Smile: Development and Application of an Instrument to Measure Product Emotions, in S.J. Hoch, R.J. Meyer (Eds.), Advances in Consumer Research, vol. 27, Provo, UT, Association for Consumer Research, 2000, pp. 111–117.

14. http://www.innerscoperesearch.com/news/Neurostandards_Press_Release_FINAL.pdf

15. http://www.emsense.com/news/electroencephalography-EEG-brainwave-measurement.php

16. http://www.neurofocus.com/pdfs/NeuroFocusExecutiveBrief_BeverageTCE.pdf

brain that is also associated with emotional response. The Advertising Research Foundation (ARF) completed an initial review of biometric and quantitative neural science measures[17] in an industry effort to help establish standards for these new consumer response techniques in research.

Choice, ranking, and other ratings (e.g. best–worst) can also be used as indirect measures of emotional response. These measures require minimal cognitive effort, by research participants making them effective when the stimuli are simple. For more complex stimuli (e.g. full concepts), the cues and emotions can be obtained by first inferring which ones are possibly at play, and then using them as the independent factors in choice or ranking models (e.g. conjoint, perceptual maps).

The uppermost level is actual consumer behavior – sensing (e.g. habits), selecting (e.g. shopping), seeking (e.g. Internet search) and sharing (e.g. peer-to-peer influencing). In the past, these metrics have been largely observational (ethnography) or self-reported through surveys. Today, there is an emergence of new techniques – enabled by the Internet. Many of these techniques are used to generate or simply tap into that information previously identified as "the river of information" (Chapter 3).

The Internet holds a wide range of new possibilities for collecting information from all four levels of the Behavior Pyramid. This includes a category of activity-based research methods such as diary and journaling techniques (as discussed earlier in the collecting of context information). These techniques are ideal in collecting information about sensing and selecting behaviors. They involve tasking engaged research participants to go out and provide content about not only what are their behaviors and experiences, but also why. This includes behaviors associated with habits such as shopping, product preparation, product consumption, and socializing. These techniques are enabled by Internet technology where content can be uploaded to help consumers tell their stories.

Sharing and seeking behavior information is also facilitated through the Internet. This includes the crowd sourcing techniques as previously discussed to characterize consumer behavior within online communities. There are a number of different types of online communities. Kennedy and Verard[18] characterize online communities on the basis of whether they are open or closed (i.e. pre-recruited) to the public, branded (specific to a brand), or independent (general research purpose). Online social networks such as Facebook, MySpace, and LinkedIn are not in themselves an online community, but have been used to recruit people into online communities or panels. Examples include P&G who uses Facebook to recruit consumers into a branded, open community for their Pringles brand (www.facebook.com/Pringles). Companies, such as

---

17. http://www.rethink.thearf.org/assets/neurostandards-homepage

18. J. Kennedy, L. Verard, Online Community Platforms: A Macro-overview and Case Study, in Online Research 2009: Online Panels and Beyond. Proceedings from ESOMAR World Research Conference. Chicago, 26–28 October, 2009, pp. 80–93

Communispace (www.communispace.com), develop custom, closed communities for brand owners. A number of marketing research companies, such as Brainjuicer (www.brainjuicer.com), have developed closed, non-brand specific online communities to generate client-specific information.

These examples reinforce the points brought up in Chapter 3 that the primary research problem today is not generating information, but how you choose to apply information sources to develop insights. The Behavior Pyramid provides a conceptual overview on how to structure the information that is collected through research. It also provides a process perspective on how to structure the building of knowledge throughout the innovation and development process.

Shelf-Life of Information

The four level structure of the Behavior Pyramid provides an additional perspective worthy of note – the shelf-life of information. The top level of the pyramid contains the most dynamic, least stable, information about consumer behaviors – sensing, selecting, seeking, and sharing. The bottom level – consumer information – is the most stable. However, there are important differences in dynamics to consider even within these four levels. Within the consumer level (bottom), identity information such as gender and personality traits remains stable throughout the life of consumers. However, the life role, life style and life values may change over time. These can have dramatic effects on the source concerns for an individual.

Each of the bottom three levels of the Behavior Pyramid includes a measure of affect with a different dynamic. Attitudes are included in the bottom level as the most persistent – i.e. they are a measure of affect that changes slowly. Moods are viewed as part of level 2 – context – and change a bit faster than attitudes throughout the life of a consumer. Emotions are the fleeting, situational-dependent measures of affect at the experience level. Emotions motivate a wide range of different behaviors that can impact the success or failure of products.

The shelf-life of information is an important consideration in three research applications. The first is when to warehouse information for future use. Information can not only get old with regard to consumer qualities such as changing attitudes, but also in the interpreted meaning of response measures that signal quality. For example, the interpreted meaning of "freshness" changed with MillerCoors' "Born On Date" campaign in 2004.[19] With the gain in popularity of microbrews, the attitudes about what constitutes freshness in beer is changing, noted by the fact that MillerCoors has now dropped its 110-day shelf-life to ensure no stale beer is drunk. The second is when consumers are engaged for lengthy periods of time within online communities. Not only

---

19. M. Barnett, Looking for a Hangover Cure, http://www.usnews.com/usnews/biztech/articles/041011/11eebeer_2.htm, 2004

might the same concerns with information warehousing apply, but also the concern that community participants behave differently than the populations they are expected to represent – behaviorally as well as in their response tendencies. The third is when designing custom longitudinal research studies. In these cases it is important to note how one experience impacts the subsequent experience. This is where the extension of the appraisal framework into a dynamic framework, as described in Chapter 3, applies.

The Silver Bullet

Characterization of research information leads to a discussion specific to emotions measurement. Everyone in our industry is looking for that silver bullet – a way to measure emotions. There is a belief held throughout the industry that if the capability to measure emotions existed, then researchers could incorporate this new source of information into their toolbox. However, it is more important for researchers to focus first on applying the appraisal framework to build models that generate insights.

The appraisal framework is based upon theory rooted in scientific investigations, showing that emotions are a temporary (fleeting) internal state of feeling, projected upon objects, people, experiences, or oneself. The accounts by "Nora" of her feelings point out the fact that emotions can arise from conceptual thought or product experiences. This is in stark contrast to moods that are persistent, undirected internal states of feeling and attitudes that are persistent, directed states of feeling. Emotions are also discrete states of feeling, as described in the topology of 22 discrete emotions.

Emotions impact our everyday lives. They help form our habits and make us "action ready" to engage in sensing, selecting, seeking, and sharing behaviors. Our humanity trains us to rapidly and intuitively recognize emotions in others as involuntary facial, bodily and/or vocal expressions.

A number of techniques have also been introduced in this chapter to indirectly measure emotions. These include the emerging fields of biometric and quantitative neural science, as well as a number of more subjective response measures. Changes in one's own emotional state can sometimes be recognized and expressed as subjective feelings. The language that consumers use in verbal conversation, peer-to-peer networking, written statements and articulated stories can all provide clues into what emotions are or have been at play. Yet, none of these information sources have proven to be a silver bullet.

None of these techniques provides a true measure of emotion. All are clues (information sources) that give us glimpses into the "implicit mind" of the consumer. The word "implicit" is used here to emphasize that emotions are formed from within the unconscious mind. This includes the unconscious "priming" of emotional states and associated senses from past experiences into our memories and the unconscious "cuing" (matching of senses to primed senses and emotions) of current situations that elicit primed emotions. Some information sources (e.g. subjective expressions of feelings) are fraught with measurement error and bias within specific situations. It is the unconscious nature of emotions – how and why they arise – that keeps us from being able to directly measure emotions in a way that leads to insights.

Emotion insight can be achieved without directly measuring emotions. The appraisal framework provides an answer to this problem by devising a thought process for the design of research for the collection of data, and the integration of this same data into insights. The key to emotion insights is to capture sufficient richness in qualitative and/or quantitative information to be able to apply the appraisal model. Every piece of information is a puzzle piece that collectively leads to insight that extends knowledge. What more value can research offer? The fact that one can never truly measure an emotion is moot. The framework itself adds sufficient knowledge to build an accurate model. The model (heuristic or statistical) leads to insights that inspire creativity or guide decision making.

This perspective is a game changer. It shifts the paradigm of thought for innovation managers and innovators. It elevates the researcher to be the game changer. It is the framework we need to develop – not more sources of measurement.

ACCELERATING LEARNING

The Behavior Pyramid shows how emotions research can become a game changer for companies. It is not the information itself that leads to this paradigm shift, but how the information leads to knowledge of strategic value for the innovation team – delivering insights that get to the "so what" seemingly missing in a lot of research. In this section, we will discuss how learning is applied to achieve efficiency in how ideas are made successful.

Holistic Product Development

Many companies still treat their innovation and development process as an assembly line. Innovation projects have a starting point – where the brand sets goals, a team is formed to generate ideas, and the team builds those ideas into a product by going through a number of stages or phases to an end point. The point is not to gain efficiency by changing the linearity of the process as much as linearity in thinking and learning.

The assembly line mentality leads to a process where the bits and pieces of a product are developed in silos – a reductionist approach. The responsibility to design and develop the bits and pieces is assigned to domain experts – i.e. innovators, such as marketers, product developers, packaging designers, engineers, and chemists – to develop. While each may sit on the innovation team, they do not develop products holistically: they do not collaborate in how all the bits and pieces are conceptually and experientially designed and developed.

This approach generates two key problems that can ultimately result in product failure, and a "black eye" for the brand. The first problem is that consumers do not experience products in bits and pieces. Holistic design and development takes into consideration the collective impact of the bits and pieces at each phase of the process. This requires the collaborative teamwork discussed in Chapter 3. By focusing on a common goal to achieve a consumer experience, the interactions between the pieces and parts of the product (designed and developed both conceptually and experientially) can be taken into account. This includes taking into account the dynamics cascade of first, second, and third "moments of truth" where consumers get to know products and build relationships with the respective brands. It also takes into consideration how consumers get to know brands through repeat use of products.

Holistic product development was first proposed as an iterative approach to food product innovation.[20] It has proven itself to achieve efficiencies in how teams work together. Whether at the early or later phases of development, the innovation team applies their collective domain expertise to continually narrow down the definition of the product to deliver the envisioned experience. A reductionist approach often leads to a final product experience that is very different from what was envisioned. The setting of an innovation strategy gives way to the development of the target opportunity. The target opportunity leads into the narrowing of scope through the definition of a product or product line concept. Conceptual design leads to a translation of concept into a product design requirement. Finally, the requirements are developed into a final product specification. This narrowing of the definition of the product can be conceptualized in schematic form, as shown in Figure 5.5.

As the definition of the product narrows, the knowledge must increase about the target consumer and how the product serves to deliver the envisioned experience. It is here that the Behavior Pyramid takes on new meaning. The lower two levels represent the knowledge that is needed to build in the early phase of the innovation process. However, by building insights that pull from information in the upper levels of the pyramid, the insights deepen.

The learning within each iteration also cycles through a process of divergence of thought, followed by a convergence. Divergent thought is inspired through research, resulting in creative activities that expand the alternatives for consideration. Convergent thinking is also guided, with sensory or marketing research, providing consumer reactions to the experience of the bits and pieces taken as a whole. It is in this cycling of divergent to convergent thinking that an opportunity exists for research to accelerate learning even more.

---

20. D. Lundahl, A Holistic Approach to New Product Development, Food Technol. 11(6) (2006) 28–32

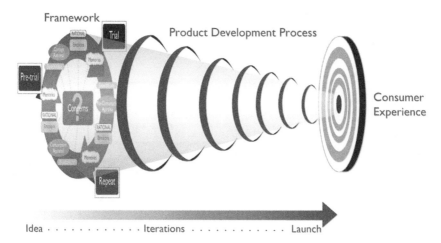

FIGURE 5.5 Holistic product development, as a highly iterative process of learning, to hit the target with a consumer experience that achieves emotional impact. (Please refer to color plate section)

Adaptive Learning

Efficiency gained through collective collaboration is enabled by a more holistic approach to research. For this reason, companies that embrace a more holistic approach to product innovation and development also tend to utilize holistic research practices to drive their learning iterations. This holistic approach to research fits especially well with the innovation approach portrayed in the "Learning Cake" (Chapter 3). However, this holistic approach can lead to an acceleration of the innovation and development process by creating closer collaboration between the researcher and innovators.

Closer collaboration can lead to an acceleration of the process by combining the principles of the "Learning Cake" with Holistic Product Development. This meshing together of the holistic, more integrated approach to research, with the holistic, more iterative approach to product development, is similar to the meshing together of two gears (see Figure 5.6).

The interconnectedness of the cogs on the left side of this figure illustrates how you can gain quality, efficiency, and speed in the development of a product. This results from taking a more holistic approach to developing and testing how each element interacts to form the consumer experience. Equally important is the process on the right side – the research learning process. This includes the relationships between study design, testing, and reporting. As we saw in the "on-the-go" case study in Chapter 3, where a new product name was created, adaptive learning occurs when you connect the learning processes on the right with the development processes on the left.

FIGURE 5.6   The interconnectedness between holistic product development (left side) and holistic research (right side). (Please refer to color plate section)

Research Efficiency

This iterative approach to learning, within the context of a more holistic approach to product development, enabled the "on-the-go" innovation team to immediately utilize insights generated within the same study. A new name was able to be both suggested and validated. Research learning that inspired divergent thinking was able to be tested within the same research study to guide convergent thinking. (See Figure 5.7.)

Adaptive learning, such as this, requires a more participatory role of researcher, innovator, and even strategist, in the process. Through rapid iteration, the team can apply their collective intelligence to more quickly generate insights. This can be in the form of adapting not only stimuli (as in this case study), but also the questions incorporated within a questionnaire or survey, moderator scripts, and other elements of the research protocols. This can lead to huge gains in research efficiency. Not only does this increase the rate of learning – leading to additional insight – it also reduces research costs. By using research participants in this way, you can effectively combine research steps, and sometimes product development steps that might require the re-recruiting of different research participants – the primary drivers of research costs.

BEHAVIOR-DRIVEN RESEARCH

Emotions research fits with this more adaptive, iterative approach to innovation. It enables innovation teams to speed up their iterative learning cycles. It leads to success by helping teams structure their information. It changes conceptual viewpoints about learning to build knowledge through emotions research from the bottom up to deliver an envisioned consumer experience.

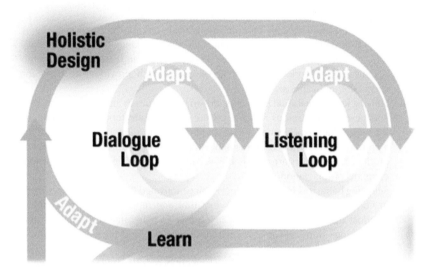

FIGURE 5.7 Adaptive research design where iterations are used to adapt the protocols for listening and dialoging, and/or to adapt stimulus within the holistic research design. (Please refer to color plate section)

This structuring of information and knowledge provides the basis for becoming a knowledge building company, experiencing huge gains in research efficiency and management of the innovation process. These sources of information provide a basis for more holistic research design, such that product design and development (conceptually and physically) can also be more holistic, taking into account the interrelationships between diverse behaviors – all elicited from a common set of emotions.

The application of informational frameworks allows for a broader understanding of connecting with consumers – on not just one or two dimensions, such as the pleasures and emotions formed from utilitarian and/or sensory appraisal – but also on a multitude of other dimensions that come into play in different contexts for use (i.e. social and emotional). These frameworks open up the innovation process to considering the importance of social factors, such as self-social identity in product design and development.

This approach to research provides the basis for behavior-driven innovation. In the chapters ahead, we will reveal through case studies how this approach to emotions research leads to gains in efficiency at every phase of innovation strategy and product development – behavior-driven innovation. With each gain in efficiency, the innovation team gains in three dramatic ways: (1) increased learning, (2) speed in defining a product that will deliver the envisioned product experience, and (3) lower costs for developing and implementing the innovation strategy.

Key Points

- Key to innovation success is for innovation teams to know how to impact the "4 S's of Consumer Behavior" — Sensing, Selecting, Seeking, and Sharing behaviors. All behaviors are motivated by emotions that can be modeled from the appraisal framework. Therefore, emotions research provides the basis for successful innovations.

- Sensing leads to habitual behaviors, driven by the cues in the environment, that engage the unconscious mind — yet are strong motivators for consumers to repeat product experiences. Understanding how to reinforce habitual use of a product — 95% of all consumer behaviors — or how to displace an incumbent product break and start new habits with a new incumbent product are key elements of innovation strategy.

- Selecting behaviors involve product choice motivated through action readiness in contexts such as shopping (product trial and repeat) and meal planning (product use). Selection behavior involves a complex interplay between the unconscious mind, driven by environmental cues, and the conscious mind making rational decisions.

- The emergence of the prosumer — seeking out brands that will work for them — suggests seeking behavior is growing in importance to the success of innovation. Seeking behaviors depend on the relative importance of choice, the accountability of the chooser, and the complexity in selecting against choice alternatives.

- Peer-to-peer sharing behavior has emerged as a huge driver of market dispersion involving influencers. As social beings there are innate factors that go beyond emotions and drive social-related behaviors. Social cues that elicit emotions and behavioral drivers that have social meaning are important differentiators for products.

- The Behavior Pyramid helps envision how to organize research information for emotions research. It portrays how information and knowledge from emotions research should be built — bottom up from the consumer to the behavior, yet always related to behavior.

- All emotions research information has a shelf life. The slowest changing are at the bottom of the Behavior Pyramid.

- There are many indirect measures of emotions to choose from. There is no silver bullet — one key emotions metric to rely upon. It is not emotions measures, but emotions insights that are important to successful innovation.

- Holistic product development changes innovation team learning from a linear to an iterative (cyclic) process with a common goal — to deliver an envisioned product experience. Companies that take a more holistic approach to product development tend also to be more holistic in their approach to research. The two processes support each other, as in the gears metaphor.

- Adaptive learning accelerates the innovation process, especially if it can be applied early in the innovation process. It more rapidly focuses innovation and speeds up the process of learning by research methods that enable the adapting of research in real time as learning occurs.

- Emotions research, as outlined in this book, leads to research efficiency in three ways: (1) increased speed in learning, (2) speed in defining a product that will deliver the envisioned product experience, and (3) lower costs for developing and implementing the innovation strategy.

Chapter 6

Strategy Development

A strategy delineates a territory in which a company seeks to be unique.

Michael Porter

BRANDS

In Chapter 5, emotions research was defined as the basis for behavior-driven innovation. This approach applies a novel, new appraisal framework – based upon the science of emotions – to generate insight into how to motivate various consumer behaviors. It is geared to achieve food product breakthroughs. When we are talking about innovation strategy, the discussion quickly shifts from products to brands ... and brands are different than products. Brands take consumers to destinations that they seek. Products provide the experience at the destination. In this way, brands are analogous to vehicles. Consumers are not going to "get on board" unless they trust the brand.

There is also a difference in affect between products and brands. Products elicit feelings as emotions, while brands elicit feelings as attitudes. Recall from Chapter 3 that emotions and attitudes have projection. However, attitudes are differentiated from emotions by exhibiting persistent states of feelings. They change slowly.

This distinction between a brand and product goes to the heart of one of the key reasons for product failure in today's marketplace. Most companies struggle to develop and implement an effective strategy that builds long term relationships with consumers. Without a trusted vehicle (i.e. the brand), consumers are not going to buy your products. Brands are built over time through repeat experiences with branded products. Yet, it is the emotional experience with a product that forms, strengthens or weakens the relationships with the brand. Therefore, it can be said that product innovation is the key to building brand relationships. However, from an innovation strategy perspective it is more important to realize that how you choose to evolve your brand to build those relationships directly impacts how you choose to innovate.

Innovation strategy provides focus to the front end of the innovation and development process. In fact, too often companies have insufficient focus at

Breakthrough Food Product Innovation Through Emotions Research. DOI: 10.1016/B978-0-12-387712-3.00006-2
87

this phase of the process. They have a very "fuzzy front end." Innovation strategy focus makes the fuzzy front end less fuzzy.

The Procter & Gamble (P&G) story, under the leadership of A.G. Lafley, provides one of the best examples of how a company is achieving success through its innovation strategy. In his book *The Game Changer,*[1] Lafley stresses three points. First of all, he defines strategy as attainable goals that are "essential building blocks for growth and should relate directly to corporate purpose that is meaningful to both employees and consumers." Secondly, he defines strategy as a "process of imagining how markets can change through innovation, and how innovation leads to value for the consumer." Third, he defines strategy as being focused on the "holistic design of the brand, the brand equity, specific product lines, unique packaging, and elements of store execution."

Innovation strategy development embraces these three "game changing" points – strategy achieves goals, strategy is a process, and strategy leads to focus. In this chapter, a process will be introduced to develop this strategy. The resulting focus will achieve clarity of vision for what will be the product experience – setting the target for innovation teams.

Finally, we will address the problem of how to respond to the consumer as a moving target. Most strategies are not focused on achieving long term business goals – while being nimble enough to adapt to consumer change. Consider, once more, the discussion from Chapter 2 for how P&G (in spite of their leadership in CPG innovation) was unable to avoid revenue losses for Tide during the recession of 2008–2009. The problem was not the recession, but the fact that P&G did not have a nimble enough innovation strategy to sense and respond to rapid consumer change. In this chapter, we will define a nimble innovation strategy that also maintains a long term vision.

THE FOUR PHASES OF INNOVATION STRATEGY DEVELOPMENT

Recall the discussion from Chapter 3 that draws an analogy between football and innovation teams. The left side of the Behavior-Driven Innovation Process is the Innovation Strategy Development Cycle, which is broken up into four phases: business strategy, brand strategy, portfolio strategy, and innovation strategy (Figure 6.1).

This analogy characterizes innovation managers as strategists, i.e. coaches. They develop winning game plans for their players (e.g. receivers and quarterbacks) on the basis of strategy. Coaches cannot develop strategy without insight into the "whys" of past performance. Therefore, strategy development requires intelligence about your team and the competition – their strengths,

---

1. A.G. Lafley and R. Charan, The Game-Changer: How You Can Drive Revenue and Profit Growth with Innovation. Random House, Inc., New York, NY, 2008, pp. 336

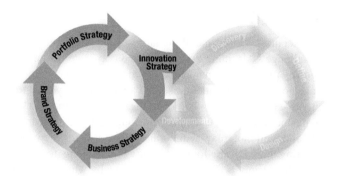

FIGURE 6.1 The Innovation Strategy Development Cycle with its four phases of Business Strategy, Brand Strategy, Portfolio Strategy, and Innovation Strategy. (Please refer to color plate section)

weaknesses, and what to expect from them in future games. In addition, strategy requires forecasts of the environmental condition at game time and foresight into how you will change your respective game plan. This analogy translates to the food industry strategist. Strategy requires insight into brand performance and the competitive landscape, and foresight into how to adapt to the changing consumer under different scenarios.

This information is essential to develop a winning strategy. It helps guide decision making as alternative strategies are considered. Strategy development is like building a foundation. Each step builds upon the former step. Early steps are more aligned with long term vision. Later steps are designed to provide nimbleness – the ability to quickly shift focus, in response to changing consumers and market conditions.

In this way, the Behavior Pyramid applies as well to strategy development as it does to product development. Targeted consumers are more apt to change their behaviors in response to new market conditions than their identities and source concerns (e.g. brand attitudes). However, if these foundations of behavior are allowed to crumble, they cannot be repaired overnight. Strategy must build long term relationships, as well as drive short term revenue.

The first three steps of strategy development iterate through the first three phases of the Strategy Development Cycle. They build the foundation upon which innovation teams set innovation strategy. The first step is to build a long term business strategy to answer the questions, "Where to play, where not to play, and how to play?" The second and third steps answer more short term questions about the brand and brand portfolio strategies, "Where are the destinations?" These destinations must be constrained to markets where a company decides to play. The understanding of where destinations are provides a foundation for a very nimble innovation strategy. This strategy answers the question, "How do we take consumers to destinations that they seek?"

BUSINESS STRATEGY

"John" is the Executive VP for one of the largest food companies in the world. His biggest challenge is deciding how to allocate corporate resources to brand teams managed by product line managers. In order for managers represented by "John" to make good decisions about resource allocation, they need to factor in two perspectives – the long term vision and the short term focus. The primary objective of business strategy is to provide long term vision for where the business will and will not play with brands. By creating new brands and managing existing brands, the food company builds long term sustainable relationships with consumers. Brands are the means by which consumers have relationships with companies.

The strategic planning process, as most companies call it in name, is really an exercise in road building. It is an investment strategy to take more consumers to more destinations that they seek. The initial investment is not into vehicles, it is into building roads. Companies must know the terrain they must navigate to be certain what kinds of roads to build. This analogy leads to a definition of the strategic planning process as a roadmap for long term business sustainability that identifies what markets to play in and not to play in (i.e. where the road needs to go), and how to play to win (i.e. how the roads need to be built to support traffic).

Road building requires vision. However, to see the right vision, the strategist needs professional surveyors – in this case, marketing researchers. So, in the sections ahead under the topic of business strategy, we will also discuss how the marketing researcher can help the strategist see the right vision for the road.

Deciding Where to Play and Where Not to Play

In 2000, P&G refocused on four core markets: fabric care, hair care, baby care, and feminine care. They made a strategic decision to exit the coffee business.[2] They invested in the development of several new brands. The result was a growth increase from ten brands with 50% of net sales and profits – to 23 brands with 66% net sales and 70% of their net profits. This success was the result of their business strategy – deciding where to play and where not to play.

Decisions of where to play, or where not to play, require foresight into how consumers and their respective markets are expected to change over time. This requires that companies continuously assess consumer and market trends – seeking to identify where existing and future brands can play, not only today, but also in the future. In the case of P&G, they monitored trends in global markets and refocused their business to play in markets that were projected to sustain long term growth. Their strategy was to play well in existing markets that had high margins and to gain a foothold in emerging markets with

---

2. Lafley and Charan, The Game-Changer

promising growth potential. Their strategy was long term focused – to invest in building new and extending existing brands to continuously build a base of loyal consumers.

This case study stresses the point that successful companies focus on playing to win through strategy that has a long term payoff. Emerging markets are but one type of market to play in. What about existing, more mature markets? Should companies exit these simply because they are not expected to grow as fast as emerging markets? The answer to these questions depends on what are the company's core business strengths and weaknesses. It depends on how a company chooses to change the game in their favor – how to play to win.

Monitoring Consumer Change

You manage what you measure. Innovation managers developing business strategy need more than insight, they need foresight. Innovation foresight is different from insight. Foresight requires the capability to turn consumer measures into forecasts sufficient to help decide where to play, where not to play, and how to play.

To address questions about where to play, many companies turn to economic data – point-of-sale data that track change in sales volume of SKUs or brands, or in growth in specific product categories or markets. However, this type of data tends to not provide innovation managers with sufficient foresight to set vision for where to play or not to play. These types of measures do not build knowledge into why SKU or brand performance is changing and/or why categories or markets are changing.

The use of economic data such as this – to set visions – is somewhat like driving while looking into the rear-view mirror. Economic data is an output, not an input to understand consumer behavior. They can be used to track, for example, "selecting behavior," but do little to understand why consumers might be changing what they select. They measure, but have no basis to forecast or predict.

Forecasting behavior requires measurable inputs into a model that is rooted in the science of consumer behavior via the appraisal framework. The Behavior Pyramid (Chapter 5) provides a basis that will guide the building of behavioral models. Consider source concerns such as attitudes about brands that were identified as Consumer Qualities at the bottom of the Behavior Pyramid. By tracking change in source concerns, such as brand attitudes among target consumers, models can be built to predict resulting changes in behavioral.

Harris Interactive[3] conduct an annual survey to measure consumer reputations of companies in the food and other industries. This survey is

---

3. http://www.harrisinteractive.com/vault/2009%20Annual%20Reputation%20Quotient%20
   Media%20Summary%20Report_public.pdf

attitudinal – measuring a number of consumer attitudes that consumers have towards companies and their respective brands. From this survey data, they develop the Annual RQ Index as a way to rank companies on the basis of perceived social responsibility, emotional appeal, financial performance, goods and products, vision and leadership, workplace environment, and being a good corporate citizen.

Companies that have a good reputation among consumers can leverage their trust for a competitive advantage. For example, at the time of this writing, Whole Foods (#18 overall) ranks #2 among those within age 40–54 – no doubt the age profile of their core consumer. Trusted consumers listen to brand-sponsored messaging and are motivated to spend their time dialoguing with brand owners. Conversely, companies that have lost trust struggle to gain the listening ear of consumers and face an uphill fight to actively engage in consumer dialogue.

McDonald's has lost some of its luster as a brand that people trust. While it maintains an active role in social responsibility (e.g. Ronald McDonald House), the concern expressed by consumers, as to the healthiness of fast food, no doubt places McDonald's at a disadvantage. It appears that McDonald's Corporation is actively engaged in innovation to change this reputation. Many of their new products and marketing campaigns are focused on health, i.e. the introduction of nutritional labeling in 2005 and more recent healthy food product introductions.

For measures of reputation and trust to make sense, consumers must have a relationship with those companies, i.e. they must know the company as a brand. For example, prior to 2008, the ConAgra Foods brand was not well developed – nor understood well by consumers. That changed with a change in corporate logo, including the corporate tag line that ConAgra Foods was "Food you love" (Figure 6.2).

Converting a company name into a brand enables the company name to be appraised – leading to emotional impact. Positive emotional impact from a corporate brand can "rub off" on all associated name brands owned by

BEFORE   AFTER

FIGURE 6.2 The "before" and "after" logos for ConAgra Foods. (Please refer to color plate section) *(Source: http://logos.wikia.com/wiki/ConAgra_Foods)*

a company – leading to attitudinal change in how all corporate brands are known as "Food you love." Therefore, a corporate brand can be used as a benchmark against which to track the reputation of each of a company's respective brands – e.g. providing information as to how well brands are living up to the promise of being "Food you love."

Attitudinal information (i.e. source concerns) alone does not provide sufficient richness to be able to decide where to play or not to play. However, by linking source concerns to demographic or simple measures of consumer identity – a new approach emerges to forecast long term opportunity for companies. In Chapter 5, the Behavior Pyramid was defined to include consumer measures of identity as distinct from source concerns. These identity measures include measures of culture, personality type, life role, life values, and traditional demographics (gender, age, socio-economic status, employment). These measures can be more easily associated with projected changes in populations and markets.

There is a growing interest in the food industry to better forecast the impact of Hispanic population growth and subsequent acculturation. By 2020, this consumer group is expected to grow to 17.8% of the US population.[4] By understanding the surface concerns of Hispanics, the impact of food attitudes, preferences, kid influence on parents, and many other factors important to innovation can be forecast.[5] By building models based upon the expected rate of acculturation of Hispanics, the change in goals, beliefs, attitudes, and ideals can be forecast, leading to important foresights for long term strategic planning.

Attitudinal research such as this helps innovation managers understand the "so whats" from research data. As with other forms of research, foresights can be deepened by going beyond simple quantitative survey research techniques. This is accomplished by engaging consumers in authentic dialogue, and listening to their comments to support and deepen foresights. Dialogue and listening information may not be trended over time through charting. However, mixing quantitative with qualitative information allows companies to understand the behavioral basis for trends, to begin to understand why various brands are or are not performing well in comparison to competitive brands or corporate benchmarks. By linking qualitative information to quantitative data, foresights can be deepened and woven into stories that inspire and guide strategic planning.

Deciding How to Play

The long term perspective essential for innovation managers to know how to allocate resources to brand and/or innovation teams comes from their

---

4. http://www.census.gov/population/www/socdemo/hispanic/hispanic_pop_presentation.html (Internet Release Date: February 08, 2008)

5. http://www.nal.usda.gov/outreach/HFood.html

knowledge of how their company and the markets they will play are changing. For this reason, it is essential that innovation managers continually update their knowledge about what has changed. This is the essence of strategic planning. A typical tactic is to conduct an annual SWOT analysis of a company's core strengths, weakness, opportunities, and threats. This analysis serves to decide how to play the game to win within the markets where the company will play.

It is much easier to focus on playing to your core strengths than trying to shore up your weaknesses. Core strengths may include aspects of your manufacturing processes, supply chain, sales and marketing capabilities, and research and innovation abilities. Strengths can also be characterized on the basis of assets – your brands, consumers loyal to those brands, and tacit knowledge – knowing how to play from past experience. Finally, strengths can also be financial – profitability leading to capital to invest.

Consider the questions posed in the last section about playing to win in highly commoditized markets. The strength of name brand owners is most likely their research and product innovation capabilities. This strength is often a weakness for retailers also playing in commoditized markets. Therefore, name brand owners can set a strategy to win in commoditized markets by innovating – changing the game in their favor by playing to their core strengths and against the weaknesses of their competition.

How a company chooses to play the game to win is controlled by how innovation managers allocate resources to brand and innovation teams. However, there are two key strategies that stand out in a company's search for breaking through the clutter through innovation. The first involves the strategy of building long term, sustainable relationships with existing consumers. The other is to be disruptive for the purpose of acquiring new relationships.

Building Sustainability

Jim Collins – historian, business consultant, and author on the subject of company sustainability and growth – defines sustainable financial performance as consistently outpacing the industry for at least 15 years.[6] Companies that outpace their competition tend to operate differently than their competition. Their goals and respective business strategy are oriented for long term sustained growth. To be – as Jim Collin defines – a great company, then your business strategy must focus on building trust and being a good corporate citizen.

Trust is a key principle for sustainable consumer relationships. If a company is a good corporate citizen, it tends also to be trusted by consumers. Why? Trust starts with how a company and its associated brands acts. Is the company authentic? Does the company strive to achieve good for the sake of society? Is it

---

6. J. Collins, Good to Great: Why Some Companies Make the Leap ... and Others Don't. HarperCollins, New York, 2001

perceived as a good corporate citizen? Do brands owned by the company deliver on their promises?

Stephen M.R. Covey spoke of the importance of trust and indicated his belief that a company's business goals need to encompass building market and societal trust.[7] Building market trust is accomplished when a company's brand(s) represents integrity or honesty – that is, for courageously addressing tough issues quickly, and honestly admitting and repairing mistakes. The brand should model good intent, or a genuine caring for people, versus being simply out to make a profit. The brand should also demonstrate capabilities. That is, the people associated with the brand are known for quality, excellence, continuous improvement, and accomplishing stated objectives. Most importantly, the brand needs to stand for achieving results and delivering on its stated promises.

Trust is measurable and, therefore, manageable. A number of marketing research sources measure trust on an ongoing basis. Consider the syndicated reports published annually by Edelman.[8] In 2009, they were able to track an unprecedented drop in trust among US companies from the recession of 2008–2009. They reported a 20-point average drop in their trust index among US businesses, including a nine-point drop among CPG companies – the greatest drop in consumer trust since the Great Depression! In addition, they reported a remarkable five-year trend of dramatic change in consumer trust, in various sources of information. Near the top of the list of sources was "consumer conversations with peers and colleagues" – nearly on par with the most trusted source "industry analyst reports and business magazines." Least trusted sources were those coming directly from a company or brand.

Edelman also tracks qualities that contribute to trust. An example is consumer perceptions about corporate sustainability practices. In 2009, Edelman reported more than two-thirds of US consumers (68%) believe American corporations are lagging behind in sustainability practices compared to other countries. Ninety-eight percent responded that it is important that corporations do evolve to more sustainable business practices, with only 16% of Americans believing that they will make these changes on their own. Authentic acts by corporations that demonstrate sustainability build relationships with consumers.

Companies that are able to build sustainable relationships with consumers have two key advantages in the marketplace. First, relationships lead to loyalty that drives consumer behavior. For example, Edelman's data showed that 90% of Americans give some consideration to sustainable business practices when purchasing a company's products and services. Stephen M.R. Covey cites a striking example from the riots following the 1992 Rodney King trial. Remarkably, McDonald's restaurants within the devastated area were untouched. The reason – consumers felt that "McDonald's cares about our community."

---

7. Stephen M.R. Covey, The Speed of Trust. Free Press, New York, 2006

8. http://www.edelman.com/trust/2009/docs/Trust_Book_Final_2.pdf

The second advantage to building trusted relationships is improved listening and dialoging. Loyal consumers will take more time to provide research information. They will be more motivated to engage in authentic conversation. They will communicate more freely with companies to whom they are loyal.

There are many authentic acts that a company can take to build sustainable relationships with consumers. This includes engaging in industry collaboration that supports efforts that are in the best interest of consumers and society as a whole. Many of these collaborations have an added corporate benefit of creating opportunity for market differentiation from innovation.

For example, many food companies are collaborating to provide improved nutritional information at the point-of-purchase.[9] This will most likely create new opportunities for companies to become differentiated on the basis of their health and wellness index scores. In the Quick Service Restaurant (QSR) industry, Subway restaurants took a leadership role in offering nutritional labels in 1997.[10] McDonald's followed suit in 2005[11] – increasing awareness among consumers and creating new opportunities for QSRs to become differentiated. In recent years, leaders in the food industry have made major strides in reducing trans-fatty acids (TFA) and, more recently, to lower salt intake. This resulted in massive efforts by food manufacturers to quickly bring to market whole portfolios of branded product lines that are still tasty, while free of TFA and with lower salt content.

These actions build corporate sustainability by building trust. A trusting, proactive consumer is seeking to communicate and take actions that demonstrate their loyalty. They are seeking to establish long term relationships with companies and brands that they trust, awarding them with a larger part of their wallet to their owned brands. Trust is a key driver of consumer behavior. It is earned, measurable, and manageable.

Being a Market Disruptor

Being a market disruptor is as important a strategy as building sustainable relationships. Disruptors focus on finding new ways to take consumers to destinations that they seek. You can disrupt markets by developing new business models, new forms of product distribution, new marketing channels, new ways to connect with consumers at an emotional level, and new manufacturing efficiencies.

Market disruptors create value for their consumers by re-energizing boring markets – using the power of imagination to consider what markets could become through innovation. Disruptors seek to differentiate brands on an

---

9. http://www.gmaonline.org/news-events/newsroom/food-and-beverage-industry-launches-nutrition-keys-front-of-pack-nutrition-/

10. http://www.thefreelibrary.com/SUBWAY%28R%29+Restaurants+Welcome+the+FDA+Menu+Labeling+Regulations-a019382297

11. http://usfoodpolicy.blogspot.com/2005/11/mcdonalds-nutrition-labels.html

emotional basis. They seek new and exciting ways to become differentiated – e.g. fighting commoditization through innovation.

Being a market disruptor is a key element of how companies choose to play the game to win. It is not just an element of strategy, it needs to be built right into the DNA of a company's culture. A market disruptor seeks to improve the performance of all their brands – not just in volumetric sales, but by making the purpose of their brands relevant to more consumers. Being a disruptor means you strive to create fulfillment through your brands – i.e. manufacturing brands with a valued purpose in the lives of a broader mix of consumers.

Collective brand fulfillment is a measurable quality of market disruption. It is measured as consumer perception for how uniquely brands fulfill important purposes in their lives. A disruptive brand changes the game – creating strong emotional connections with people – creating strong desire for repeat use.

Being disruptive has multiple benefits. First, disruptors are successful at sensing changes in consumer behavior and using that change to their advantage. For example, at a time when the laundry detergent industry was fully commoditized – consumers seeking the lowest price alternative – the Tide brand came up with liquid detergents that revolutionized the market and created new excitement.[12] Tide not only fulfilled a convenience purpose, it also fulfilled a purpose of intrigue and stimulation through its "newness." Secondly, disruptors become known in the industry as innovators. As the Tide brand revolutionized the laundry detergent category, P&G became known as a market innovator both within the industry, and by consumers. This contributed to excitement internally at P&G – and further strengthened their culture of disruptive innovation.

BRAND STRATEGY

Business strategy serves to achieve vision, brand strategy, and focus. Brand strategy focuses innovation teams on the destinations being sought within target markets. While brand strategy must align with a company's vision, its primary function is to link the product (i.e. the destination) to its brand (i.e. the vehicle). Brand strategy also involves determining how to build relationships with consumers so they will "get on board."

This point of view is in stark contrast to a focus on products to grow your business. Companies that are product driven tend to think about the brand as an attribute of the product, rather than the other way around. This leads to confusion among marketers, designers, and developers as to how best manage a company's portfolio of products. Further, it takes the focus off of building consumer relationships and how to serve the needs of consumers.

In 1993, Brown Forman Corporation test marketed Jack Daniels Beer. The product was an excellent macro-brew at a time when consumers were shifting

---

12. Lafley and Charan, The Game-Changer

their tastes from traditional large brands in the brewing industry.[13] The product failed because it did not serve to develop relationships between consumers. The product was simply a novelty – offering intrigue at first, but not impacting consumers on an emotional basis in the beer category. It did not serve the needs of the community of Jack Daniels loyalists. While loyal consumers trusted Jack Daniels, it simply did not fit with the consumer defined purpose of Jack Daniels. While consumers at the time were starting to switch to more specialty brews, the Jack Daniel's brand was not viewed as an appropriate vehicle to take them to the specialty beer destination.

Destinations

The Jack Daniels Beer case study points out two important goals for a brand – a brand must serve a purpose in the lives of consumers, and it must stand out from the clutter of other brands. In the case of Jack Daniels Beer, the brand was well known, but could not fulfill the envisioned purpose. Recall A.G. Lafley's points in Chapter 2 when he defined a key step in the P&G innovation process, to imagine the possibilities for the consumer experience. The problem is that not all brands can serve as the vehicle to take consumers to the imagined "destination themes."

Marc Gobé,[14] in his book *Emotional Branding*, discusses at length a brand's goals. He states, that for brands to be successful, they must be able to shift their purpose from "notoriety to aspiration" where "notoriety gets you known, not loved or desired." Howard Shultz, CEO for Starbucks, in his autobiography talks about brands needing to be able to "romance the consumer".[15] Put into the appraisal framework, to get to desire, the brand must convey something aligning with the consumer's aspirations – their source concerns – but cannot forget the purpose for which the brand is already notorious. In the case of Jack Daniels beer, there were disconnects in the mind of consumers between the purpose of a Tennessee whiskey and a specialty beer.

Understanding the purpose, identity, and personality for a brand are essential elements in discovering what destination themes a brand can fulfill through its line of products. To envision the possibilities, the innovation team must focus on the range of possible experiences, placing lower focus on functionality and more on feel. Many marketers design for maximum functionality. However, brands are not going to cut through the clutter with breakthrough products unless they have an identity and personality that stands out from other brands, and a purpose that takes consumers to theme destinations rooted to their fundamental lifestyles, values, and identity of self.

---

13. Personal account – having served as a member of the innovation team from 1992 to 1993

14. M. Gobé, S. Zyman, Emotional Branding. Allworth Press. New York, NY, 2001

15. H. Shultz, Pour Your Heart Into It. New York, Hyperion, 1997, p. 5

Brand Goals

In hindsight, it may seem as if the brand team for Jack Daniels made a poor decision in trying to launch a line of specialty beers. Yet, it all depends on the goals of the team in exploring this innovation. Brand goals fall into two categories: offensive and defensive. Defensive goals seek to protect the brand. This can involve reinforcing habits that make it more difficult for competitors to disrupt the unconscious mind. This often happens with brands that have established themselves as a category leader, or for a new brand seeking to protect its unique position in the marketplace. However, just like in football, brand teams cannot rely on defense alone to win.

Offensive brand goals seek to grow their line of products – i.e. they seek to take the consumer to a wider range of destinations. They establish goals to growing the product categories they play in, or to taking market share away from competitors. Offensive goals serve to develop, extend or expand the purpose of a brand. Brand development serves to differentiate your brand from your competition among an existing base of category users. Brand extension serves to acquire new consumers, by discovering new ways to increase the context of use or to find new opportunities for the brand to extend into new product categories. Brand expansion broadens the purpose of a brand by endorsing new brands with a different, but related purpose. This includes goals to leverage existing brand loyalty and equity to grow into new product categories and business areas.

The final type of brand goal is to create a completely new brand that is not endorsed by an existing brand. This typically occurs when a company focuses innovation on the creation of a truly new product – for the company, for a target market or for the world. In the case of Jack Daniels beer, the brand team was taking an offensive strategy – seeking to leverage the brand's loyalty to enter into the specialty beer market.

Brand KPIs

Performance against goals that can be measured can be managed. Many innovation managers set key performance metrics indicators (KPIs) to gauge the effectiveness of brand teams. It is typical for brand teams to monitor brand performance on the basis of sales volumetric metrics from point-of-sale (POS) data. However, as was noted in the earlier section on monitoring consumer change, economic measures typically are poor to help forecast or deepen insights into the "whys" of brand performance. The appraisal framework and respective information within the lower levels of the Behavior Pyramid™ provide a wealth of better alternatives.

In Chapter 5, the technique of sentiment analysis was introduced as a way to capture and analyze attitudes toward brands from online peer-to-peer chatting within online communities. This technique is being applied to track the performance of brands (e.g. Conversiton[16]). By sampling from the constant flow of the online community, the peer-to-peer dialogue, general or target population attitudes can be continually tracked about specific brands. Changes in attitudes to the brand can be tracked to help assess the impact on the brand from new product introductions, marketing campaigns, and external market factors.

Which alternatives are best as brand KPIs depends in part on brand goals. If brand goals are defensive, then a better alternative might be to measure emotional impact from brand appraisals among loyal consumers; or, to track attitudes among target consumers. Alternative KPIs associated with offensive goals might be measures of gap in emotional impact on brands vs. their chief competitor, or to track changes in attitudes among a brand and its competitors. The attitudes chosen would be those shown to ultimately impact sales volume, leading to insight and learning upon which the team might shift brand strategy in response to consumer change.

Brand KPIs provide a basis against which teams can not only assess the impact of strategic decisions, but more importantly provide a means to build strategy.

Developing Your Brand Strategy

Once goals and KPIs are in place for a brand, the next step is to use these to develop a brand strategy. Brand strategy is a means to achieve brand goals – e.g. a way to enter into a target market by leveraging the assets of a brand. When performance against goals can be measured as KPIs, they provide a basis to understand the types of destinations that might be appropriate for brands.

A brand team that tracks how consumers perceive the purpose, identity, and personality of its brand as KPIs can use this knowledge to critically assess new destinations to extend or expand its purpose. If the Jack Daniels brand team had been able to map out the brand's purpose, identity, and personality, it may have seen there was a misfit with the purpose of a brand associated with Tennessee Whiskey and a specialty beer.

In this way, brand strategy should be much more than simply a plan to extend its product line. It is a behavioral argument on how a given goal is to be implemented and why it will be successful. For example, if the goal of a brand is to protect itself from competition, then a strategy might be to achieve the goal by building loyalty – creating action readiness to maintain (reinforce) use habits. A goal to develop a brand requires a strategy to increase brand share – disrupting the unconscious minds among users of competitive brands to get them to try to switch to a new alternative.

---

16. http://www.conversition.com/evolisten/

This strategy of evolving brands against goals is a departure from the typical strategy – a strategy focused on marketing to continually build awareness. It takes into consideration how to change behaviors by applying the appraisal framework through innovation. This includes a broader mix of contributors to innovation – not only marketers, but also researchers, designers and developers. Instead of the brand focusing on how to differentiate new and existing products, the focus of the brand is placed on the consumer and what motivates consumers to become action ready.

The Kettle® Brand Potato Chips Case Study

Kettle Brand Potato Chips, by Kettle Foods, developed its brand from a home-based business in 1980 into the leading brand for "all natural" potato chips. They were acquired by Diamond Foods in 2009. In an interview with Chief Ambassador, Jim Green,[17] he described how they set and achieved their goal to develop their brand. In 2004, they set a goal to build and expand their community of brand loyalists. They developed an innovative approach to achieving this goal by engaging their most loyal consumers — asking their opinions for how to develop products that the community might most desire. Their strategy included the launch of their "Create-A-Chip Challenge," soliciting ideas, through their website, for flavor combinations to extend their product line. They even offered a Create-A-Chip Kit for $15, with seven bags of all natural seasonings, so participants could experiment to come up with new ideas. Thousand of ideas were entered and synthesized into five categories with choices for voting. The winner — "Spicy Thai" — was subsequently developed and launched into the marketplace. This process has been subsequently refined and used to develop successful products such as "Death Valley Chipotle." As a result of these efforts, Kettle Foods has been able to achieve more than simply developing its brand, it has deepened relationships with its consumer base.

The purpose of the Kettle Brand has evolved to be more than just an all natural tasty snack, but to help consumers feel that they were part of a community that supported their beliefs and lifestyle. Each product in the different lines under the Kettle Brand might appeal to different tastes and/or serve a different purpose. However, the ideals, attitudes, beliefs, and goals of the whole community include support for sustainability practices associated with "natural and organic" foods.

By understanding a brand's purpose, and the purpose of competitive brands, a stronger strategy can be envisioned for how to evolve the brand. This does not necessarily imply a strategy to change the brand's purpose, but to evolve how different consumers find value in that purpose. As a result, the actions of the Kettle Foods (and the Kettle Brand) loyal consumers increased their desire to support the brand (and their own ideals, beliefs, attitudes, and goals) and to attract new consumers into the community who were seekers hoping to find a brand that supported their held ideals, attitudes, beliefs, and goals.

---

17. Personal interview with Jim Green, August 2010

BRAND PORTFOLIO STRATEGY

The establishing of a brand strategy leads naturally from questions of what is the destination to how best to deliver the consumer to a destination. This starts with an assessment of the product portfolio for the brand and how it should change. How might a brand leverage its current portfolio of products – and product lines?

A brand's portfolio includes the products managed by the brand team – the products that fit under the brand umbrella. The brand goals help identify why the brand team chooses to evolve a brand. The brand strategy is a translation of this goal into more behavioral or consumer terms, i.e. how your brand goals will be achieved by applying the appraisal framework. The brand portfolio strategy uses brand strategy to focus innovation on evolving the portfolio of products under a brand.

Consider Table 6.1, which relates brand goals and strategies to how the portfolio might evolve. This table shows that the brand portfolio strategy falls naturally from a brand's strategy to achieving its respective goals.

TABLE 6.1 Brand Goals and Associated Strategies

Brand Goals	Brand Strategy	Brand Portfolio Strategy
Protect your brand	Increase brand loyality through deeper emotional impact	Create competitive barriers for how a brand fulfills a job critical through product line optimization
Develop your brand	Increase brand share through deeper emotional impact	Differentiate product line against competition on how well branded products fulfill a job
Extend your brand	Expand the emotional impact and REACH of a brand into new markets	Finding new markets for a brand to fulfill its job by developing new product lines; for new consumers with a different market definition for a job, for new use contexts (more occasions), or for new categories (different types of occasions)
Evolve your brand	Leverage the emotional impact and REACH of existing brands to create new brands with greater emotional impact REACH	Create sub-brands that are "endorsed" by a brand with products that fulfill different jobs for existing consumers of the brand (new types of trials)
Create a new brand	Achieve new ways for emotional impact & emotive REACH through a new brand	Create new brands with products that fulfill job(s) within a new product category or with new consumers (new trials)

If the brand strategy is to increase loyalty to protect the brand, each product line must be managed to create competitive barriers for how the brand fulfills source concerns in unique (protectable) ways. If, on the other hand, the brand strategy is to develop the brand by increasing share of a market, the brand portfolio strategy shifts to offense – seeking differentiation among existing consumers for how well a brand fulfills a job. Strategy to extend the brand involves the development of a plan to manage a brand portfolio for engaging new consumers in different markets, to identify new use contexts within which the brand serves its purpose, and to find applications in different categories for the brand to compete by getting the same job done. When creating new brands, the goal is to achieve broader emotional impact and reach with sub-brands that fulfill jobs within some new product category for new consumers.

These possible courses of action for the brand require that the line of products, within the portfolio, are optimized to achieve emotional impact among existing consumers, and to broaden a brand's reach within a market or into different markets. Extending a brand's reach is about expanding emotional impact, by discovering new markets for a brand to fulfill its job. It could be identifying new consumers, trials, use context, occasions, or categories.

Expanding brands leverages the existing emotional connection with consumers to create opportunities for new brands or sub-brands to be manufactured – extending the reach of brands beyond their purpose through an endorsement. Examples of endorser brands include General Mills endorsement of Cheerios and ConAgra Foods of Healthy Choice. They each have leveraged their brand into a wide range of different markets – getting a multitude of different jobs done by leveraging the connectivity each brand has to consumers on a global basis.

INNOVATION STRATEGY

The final step in the process of developing an innovation strategy is to ensure there is sufficient focus for the front end of the innovation process. This involves decisions as to which type of innovation fits with the risk–reward acceptable to the brand. It also leads to nimbleness.

Types of Innovation

The selection of the right type of innovation is essential to how you approach the front end of the innovation and development process. Christenson and Raynor[18] characterize innovation in two different ways. The first is on the basis of how innovation arises within the organization. Is it emergent or deliberate?

---

18. C. Christensen, M. Raynor, The Innovator's Solution: Creating and Sustaining Successful Growth. Harvard Business School Press, Boston, MA, 2003, pp. 304

Emergent strategies are tactical, day-to-day focused, and ideal for uncertain markets, such as we have today. The focus is not far into the future and is all about incremental, tactical movements that are not necessarily planned. Deliberate strategies are top down-driven, purposeful, and analytical. They require a purposeful intentionality that focuses innovation on the basis of information external to the innovation process.

A deliberate approach is appropriate when the market opportunity is clearly identified or when there exists an innovation "platform," i.e. a technology platform exists that will achieve a protectable competitive advantage. Food retailers tend to use deliberate strategies, as they are copying successful new products already launched by name brands. Café Steamers was an example of a deliberate strategy around a packaging patent. However, when there is a great deal of uncertainty in the marketplace, it is better to shift to a more emergent strategy.

Christenson and Raynor also characterize innovation on the basis of how they impact markets. Is an innovation incremental or disruptive? Incremental strategies tend to be used when trying to protect or slowly develop a brand – yet do little to expand or extend a brand into new market opportunities. An example is a strategy to optimize product lines for cost reduction. These innovation efforts strive to squeeze more profit out of products without negatively impacting the consumer experience and brand. Another example of an incremental strategy is one that focuses on adding flankers that extend product lines such that they protect the brand from competitive threats. Incremental strategies may increase bottom line profitability in the short run, but do little to achieve long term brand goals.

Disruptive strategies tend to be deliberate attempts to disrupt markets, as discussed earlier in this chapter. Breakthrough food product innovations tend to be disruptive. They take market share away from competitors by eliciting emotional impact – in highly differentiated ways. They often lead to the formation of entirely new brands. When they are used in conjunction with existing brands, they tend to greatly extend or evolve a brand's purpose. These strategies extend a brand into new categories, new situational contexts of use, and/or new consumers.

Disruptive, deliberate innovation tends to have the higher risks and rewards. An important consideration is to properly balance risk–reward. The lowest risk involves incremental strategies that are geared to simply sustain (protect) brands. Likewise, the highest risk involves developing and launching new products into categories where your company has no or little experience, or to attempt to develop your own category through an innovation (see Table 6.2).

Thomas Kuczmarski,[19] in his book *Managing New Products*, discusses various approaches to manage risk–reward. Risk tends to increase when going

---

19. T. Kuczamarski, Managing New Products. Book Ends Publishing, Chicago, IL, 2000, pp. 274

TABLE 6.2 Innovation Strategies and Associated Risk

Brand Goals	Innovation Strategy
Protect your brand	Sustaining: incremental innovation for product improvement, line extensions and/or repositioning through innovation that improves experiences for existing consumers
Develop your brand	Expansive: incremental innovation for product improvement, line extensions and/or repositioning by focusing innovation on new experiences for existing consumers
Extend your brand	Expansive: incremental or disruptive innovation for developing new lines and products for new markets under existing brands by focusing innovation on new experiences for new consumers and/or new use contexts for existing consumers
Evolve your brand	Expansive: incremental or disruptive innovation to create new sub-brands, lines and products for new markets by focusing innovation on creating new experiences
Create a new brand	Expansive: disruptive innovation to develop new brand, line and products for new markets by focusing innovation on creating new experiences

Increasing Risk

from a sustaining to an expansive innovation strategy, or from incremental to disruptive. Therefore, you can mitigate your risk by maintaining a sustaining and incremental strategy. Risk–reward can also be managed by adopting a business strategy to have a mix of low to high risk innovations, developing risky new products for some brands and safer line extensions for others.

Yet, there is an innovation type that neither Christensen and Raynor nor Kuczmarski discuss. This characterizes innovation on the basis of its flexibility. Is it nimble (very flexible) or rigid? The four-phase approach to developing innovation strategy enables any of these forms of innovation to evolve. The business strategy is more deliberate and disruptive – staying the course to achieve a long term vision with a more disruptive market strategy. The brand,

brand portfolio strategies and innovation strategy itself become incrementally more flexible as the process of innovation strategy development proceeds to the front end of the product development process.

The enabler of this strategy is the behavioral approach taken in this book. As the process moves from the development of business to innovation strategy, risk is continually mitigated through learning. By being more behavior-driven in the development of business strategy, a company can identify markets with the highest long term reward. Brand teams can also reduce their risk by understanding how to identify destinations that fit with or extend a brand's purpose. Finally, the innovation teams can collaborate to identify whether an emergent or deliberate, disruptive or incremental strategy fits the market situation and degree of risk–reward at play.

The Healthy Choice Case Study

Consider the example of ConAgra's Healthy Choice. In 2009, this brand had a product portfolio organized into four product lines or sub-brands: all natural ingredients; "Café Steamers," with a packaging technology that "unlocks" fresh taste by steam; complete meals that provide the essential nutrients; and "Select Entrées" developed by world-class chefs. The baseline requirements for this product category are to be single serve, convenient (frozen storage and microwavable), and have an appealing taste. Each one of these sub-brands is focused on appealing to a different market segment, has a different meaning for what is "Healthy Choice," and a different definition of what is a "healthy eating" job-to-be-done.

Healthy Choice competitive brands include Smart Ones and Lean Cuisine. All three competing brands appear to target "healthy choosers" in the given contexts of use and fulfill the baseline category requirements. Smart Ones is associated with Weight Watchers and touts a "bistro taste." Lean Cuisine is labeled "café fresh, no preservatives and low fat." Each of these brands and the four Healthy Choice lines is differentiated from its competition on the basis of a different set of product jobs.

Consider the "Café Steamers" line. This sub-brand fits well within the Healthy Choice brand provided consumers associate "steamers" with healthiness, i.e. "steamed with a fresh taste unlocked by steam." It is differentiated among consumers who accept the claim that "steamed unlocks fresh taste." The product experiences of microwave cooking and consuming products from this product line are the result of a disruptive packaging innovation — a patented technology for microwave steam cooking. Product positioning and claims also contribute to these experiences by establishing expectations in the mind of the consumer.

"Café Steamers" was the latest line introduction to Healthy Choice. The question about how to evolve the Healthy Choice brand required first a decision of why to evolve this brand. In this case, it appears that the Café Steamers line was added to develop the brand — i.e. to take market share away from competitors, rather than to grow the category.

The decision to add a new product line such as Café Steamers was costly and risky. Successful innovations, such as this, require that an innovation strategy be based upon insight into how the marketplace is changing – how consumers define the meaning of a brand. Brand maps provide a structure upon which to base your decision as to how to focus innovation – should you simply optimize existing lines or develop a whole new line to achieve your brand goals? This requires insights into how consumers segment on the basis not just of goals and "ought" (e.g. to eat healthy), but also on the basis of attitudes and ideals (i.e. what is healthy eating or the ideal healthy entrée).

BEHAVIOR-DRIVEN INNOVATION STRATEGY

The behavior-driven approach to innovation stresses the importance of strategy development to set up the product development process for success. It involves a four-step process that is deliberate to achieve long term business goals and is nimble to achieve more short term goals for brands through product innovation. The payoff for strategy development is greater focus at the front end of the product development process for innovation teams. This leads to greater learning, speed in getting to market, and success in achieving product breakthroughs.

Nimbleness is a key outcome of behavior-driven innovation. It solves the problem for food companies seeking to keep pace with the ever-changing consumer, without losing sight of longer term, more strategic business goals – where and how to play, and how to play the game to win. Through emotions research, a brand strategy can be developed to focus innovation of how to take consumers to destinations they seek. At the later phases of innovation strategy development, this approach to research enables innovation teams to pick the best strategy that mitigates the risk–reward.

Finally, this process for innovation strategy development is represented as a cycle. The cycle of innovation strategy development starts and ends with another cycle of strategy planning. Each phase serves to build upon the knowledge from the preceding phase in an ongoing process of learning through research. This sets the stage for a more successful product development process.

Key Points

- The acceleration of consumer change requires companies to keep a finger on the pulse of change in the destinations that consumers seek, and to become more nimble in their speed of response through innovation.
- A nimble innovation strategy starts by thinking behaviorally. Behavior frameworks apply our understanding of human nature to better anticipate and sense consumer change. Nimbleness is an operational response to consumer change by building desired destinations for consumers through innovation.
- Your innovation strategy can be developed using a four-step process: establish a sustainable business strategy for corporate growth (step 1); use your

understanding of consumer change to develop strategy to evolve brands (step 2); evolve brands by developing strategy for how to evolve product lines (step 3); and use steps 1—3 to establish an innovation strategy that achieves focus for the front end of the product development process (step 4).

- A sustainable business strategy involves deciding on where to play and not to play, and how to play the game. This includes focusing on the goal to build sustainable relationships with consumers in markets where you have decided to play. Critical to building sustainable relationships is the earning of trust amongst those consumers through corporate actions that prove your authenticity to work for them.

- A successful brand strategy builds sustainable relationships with consumers by demonstrating through action that the brand works for them.

- The definition of brand success should involve more than financial, but also key performance metrics associated with a brand's reach in building sustainable relationships with consumers.

- Brand relationships are built by sensing change and using this knowledge to evolve consumer attitudes about the purpose the brand is to fulfill.

- Established brands are evolved through strategy to develop a brand's portfolio of products (i.e. product lines and sub-brands) in response to consumer change that maintains and evolves consumer attitudes about the brand.

- Behavior-driven innovation strategies focus on the front end of the product development process, i.e. they make the fuzzy front end less fuzzy.

- Focus involves identifying the right type of innovation and strategic arena that aligns with your business, brand, and brand portfolio strategy.

Chapter 7

Discovery

Discovery consists of seeing what everybody has seen and thinking what nobody has thought.

Albert Szent-Gyorgyi (1893–1986), in Irving Good, *The Scientist Speculates* (1962)

THE "FOCUSED FRONT END"

Behavior-driven discovery can best be characterized as a highly-focused front end to the product development process. In Chapter 3, behavior-driven innovation was introduced as two interconnected learning cycles – innovation strategy development and product development. Processes, methods, and techniques for innovation strategy development were introduced in Chapter 6 – providing focus to the front end of product development. Innovation teams use the innovation strategy to focus on domains of consumer experience to discover the best ideas for how to evolve a brand. This focus leads to an acceleration of the product development process.

The speeding up of product development through focus is different from the speed gained through more efficient emotions research, as discussed in Chapter 5. Product development speed is often equated with the collapsing of or cutting out of phases. This is not the case with the behavior-driven approach!

The output of innovation strategy development serves as the input for discovery; this includes knowledge of what is the brand, destination, line, and innovation type. This strategy includes inputs as to what risk–reward balance to accept and will include decisions as to whether the innovation team will deliberately focus innovation on a given marketing or technology platform. It will have identified whether the objective is to achieve a breakthrough innovation that disrupts markets, or to take a very incremental, low risk approach, and develop a line extension or flanker.

In behavior-driven discovery, these decisions would not have been made in a vacuum. In fact, the research used in developing this strategy also serves as a knowledge base to build upon in the Discovery phase. This chapter will introduce research methods and techniques that extend this knowledge. In addition, techniques will be introduced to apply this knowledge to discover the biggest opportunity to take consumers to the destination that they seek.

Breakthrough Food Product Innovation Through Emotions Research. DOI: 10.1016/B978-0-12-387712-3.00007-4
109

These new methods will be presented in the form of stories and case studies to help visualize how these applications can be used to make innovation teams be more successful. The pay-off of this chapter will be the arming of innovation teams with an arsenal of new research tools, methods, and approaches to accelerate and make more effective the discovery process.

DISCOVERY – THE FRONT END OF INNOVATION AND DEVELOPMENT

Discovery is one of the least streamlined and standardized process steps for innovation in the food industry. Part of the issue is a lack of standardization throughout the food industry. Some companies blur the definition between innovation and product development. Other companies draw solid lines to distinguish these two – i.e. innovation only pertains to new products or the strategic planning prior to discovery.

In Chapter 3, the idea of a more holistic definition of a food product was introduced. In Chapter 5, this idea was extended to define holistic product development different from the scientific, reductionist approach taken by many food scientists in conducting research. This holistic definition of a product leads to an expanded definition of product development. It includes all activities that take an innovation and development project from idea to launch. This requires that product development activities include those that lead to a definition of the opportunity, concept, product requirements, product specifications, and marketing mix. Whether incremental or disruptive, emergent or deliberate, this broader definition leads to the necessity for every product development project to iterate through a standardized process.

Product Development Cycle

Discovery is the front end of a four-phase process called the Product Development Cycle. This is the right side of the Behavior-Driven Innovation Process, as defined earlier, in Chapter 3. The four phases of a Product Development Cycle are Discovery, Define (i.e. concept design), Design (product), and Development (Figure 7.1). What distinguishes this process from traditional product development processes (e.g. Stage-Gate®[1]) is that their focus is on how innovation teams learn and apply their learning. This process is viewed as a cycle, with inputs from the back end of the Strategy Development Cycle and from the Development Phase. The Product Development Cycle can easily fit into a number of traditional product development processes. For example, the Discovery, Define, Design, and Development phases can easily be implemented within the gates required for Stage-Gate or other phase-gate processes. One can

---

1. R.G. Cooper, Winning at New Products: Accelerating the Process from Idea to Launch. Perseus Books Group, Cambridge, MA, 2001, pp 446

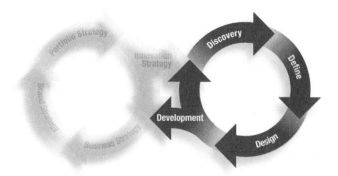

FIGURE 7.1 The Product Development Cycle with its four phases: Discovery, Define, Design, and Development. (Please refer to color plate section)

put a gate in between any of these phases as management control points to ensure that the goals of a project are being met – and to manage the funding of projects that become more and more costly at the back end.

Discovery is the process by which opportunities are identified in the marketplace and converted into ideas that have distinct market potential. The goals for Discovery and subsequent phases must be aligned with the developed innovation strategy. A more disruptive, high risk innovation strategy is expected to require a much more detailed and time-consuming discovery process than, for example, a short-sighted, low-risk incremental innovation strategy.

Consider the case of a productivity enhancement to a product line to increase revenue margins for the brand. There is very little discovery, scoping and design required to implement this strategy. Most of the work will be in the development of alternative suppliers and business processes for the manufacture, distribution, and shelf placement of products at lower costs.

The process is distinctly different in the case of a product development project to extend the brand into new situational contexts for use. Here the Discovery process is extremely critical for innovation success. It would require that much more be learned about current contexts for use, what jobs are being done by products under the brand, and what unfulfilled jobs – aligned with the brand's purpose – are currently underserved by the brand and its competitors. This Discovery process requires a different set of goals, research plans, research techniques, and learning processes because the focus is different.

DISCOVERY GOALS

The goal for the Discovery process is to identify the biggest opportunity to impact the behaviors of target consumers. The goal depends on the types of behavior that the brand chooses to alter or impact. These target behaviors form the guardrails within which the innovation team focuses its creativity and decision making to discover the biggest opportunity for the brand.

These goals are achieved through the development of product experience themes that characterize the destinations that consumers seek. These themes result in three types of behavior change: (1) attracting new consumers to the brand; (2) inspiring new use cases for products; and (3) breaking old and establishing new habits. Each forms a different type of goal to be achieved through the Discovery process.

New Consumers

The first type of goal is to attract new consumers to the brand. These new consumers seek a common destination through an existing, improved or new product experience. This type of opportunity tends to be identified through research involving data and knowledge from the lowest level of the Behavior Pyramid™ (Chapter 5).

This information framework provides the basis for understanding how to focus the discovery process to identify consumers with similar behavioral tendencies. At the top of the pyramid is information about actual consumer behaviors, e.g. what products consumers have purchased, considered for use, prepared for use, or actually used through some consumptive behavior. More often than not, segmentation research is based upon surveys probing into self-reported habits (e.g. brand usage). Other times it is based upon observed choices between experimentally controlled alternatives or comparative assessments (e.g. feature or taste preference). Self-reported habits, choice, and, to a lesser degree, preference all depend on specific context. Consumer habits tend to depend on when, where, and what we are trying to get done through usage. The choice alternative will often change with context. Feature or taste preferences will vary in ranked importance from situation to situation. Even ethnography and other observational techniques depend on the context of situations.

Segmentation research based upon behavioral tendencies is different. It focuses on the bottom of the Behavior Pyramid – focusing on identifying population qualities on a behavioral basis. This provides a more solid foundation for the innovation team to generate insights into target consumers seeking out a common destination. The bottom of the Behavior Pyramid includes qualities of consumer identity, as has been discussed earlier – i.e. their demographics, life stage, life roles, values, culture, and social economic status. However, these qualities by themselves are poor predictors of behavioral tendencies.

The identification of population qualities, on a behavioral basis, starts by collecting information about the similarities and dissimilarities of consumer source concerns. As defined in Chapters 4 and 5, source concerns are the goals, standards, attitudes, or ideals that consumers use to direct their general tendencies in taking action. The term "general tendencies" refers to the behaviors that consumers will tend to take across a range of specific situational contexts.

For example, consumers with a common goal to lose weight will generally have similar behavioral tendencies. They may frequent similar situational contexts such as health clubs or more "health-food" oriented retail stores. Their source concerns about achieving weight loss will lead to similar surface concerns in specific situational contexts. As a result of having similar source concerns, they will tend to respond similarly – having similar surface concerns at a point of experience and ultimately experiencing similar emotions.

Source concerns not only can be causally related to behavioral tendencies, they often associate well with more fundamental consumer qualities such as gender, lifestyle, and life values. Consumers seeking to lose weight may be associated with specific demographics such as age classes and gender.

The discovery of consumers with similar source concerns goes beyond consumer goals such as losing weight. It can include the discovery of market segments with common beliefs in what behaviors ought or ought not to be done. These "oughts" are often shaped by values, cultural-specific norms, and the stories that consumers keep about their self-social identity. Attitudes are entirely different, yet equally important in shaping behavioral tendencies. As noted in Chapter 4, attitudes have projection-like emotions, yet are persistent (more akin to moods). Attitudes can be projected onto oneself, onto other people, or even onto brands and/or entities such as companies that own brands. Attitudes can also be projected onto experiences or situational contexts. Attitudes – like emotions – impact behavioral tendencies by attracting or repelling consumers from brands, brand owners, contexts and specific product experiences. Discovery can also be based upon characterizing behavioral tendencies on the basis of common ideals. Ideals shape preferences, providing a basis for benchmarking what is or is not good enough.

It is through the generation of research information into behavioral tendencies that behavioral insights are formed, leading to the discovery of new target consumers. By characterizing consumers with common source concerns, a more solid foundation is established for innovation teams to attract new consumers to the brand – i.e. to identify consumers seeking a common destination. The persona of such discovered market segments can be defined through a story board with an identity, personality, and source concerns. Through these sorts of stories, researchers and marketers come to know market segments and their common behavioral tendencies. When told, these segments come to life, allowing the innovation process to achieve focus.

New Contexts for Use

The characterization of new or existing segments of consumers forms the basis for further opportunities for the discovery of new use cases for products – i.e. to identify contexts where products can fulfill unfulfilled jobs-to-be-done. Consumers with similar behavioral tendencies may or may not share similar contexts for purchasing, preparing to use, or using products. However, behavioral tendency segmentation provides a basis for discovering the jobs that consumers seek to fulfill in different contexts. It is through insights into unfulfilled jobs by these market segments that context opportunities are identified.

Context is the next level in the Behavior Pyramid. This level characterizes qualities of situational context within which consumers encounter products. It can be segmented into three distinct contexts: purchase, preparation to use/consume, and use/consumption. The purchase context is important when consumer insights are being sought to discover opportunities for retail inno- vations, i.e. package or merchandizing insights. This might be in the discovery of new context opportunities for engaging with consumers or attracting their attention. The preparation-to-use or consume context is important when consumer insights are being sought to discover opportunities for package or preparation-to-use innovations. The third type of context, the context of use or consumption, is often the same as the context for preparation – but not always. An example is the context for applying a facial makeup product (e.g. using an applier) and the context of use (e.g. where the product is worn).

This type of behavioral information includes qualities that characterize the context of situations (environmental and consumer state of mind), and the jobs- to-be-done by products for consumers in that context.

Behavioral information about situational context includes the aspects of environment that generate sensory cues that prime consumers for action. Consider a consumer at home watching TV: in that context, an advertisement that portrays a consumer eating a chocolate bar may provide cues that prime the consumer for going to the pantry and pulling out a chocolate bar or, if there is none available, making a mental note to purchase one the next time the consumer is at the store. Environmental cues can also be completely subcon- scious – such as the smell of a hamburger at a fast food restaurant that cues a consumer to order the double cheeseburger when they originally intended to order the low-fat fish sandwich. Environmental elements of context that impact behavior also include factors that cue our habitual behaviors, such as time of day and place of encounter (e.g. at home, in the kitchen or bathroom, in the car, at work). They also may characterize our normal routines (e.g. brushing teeth) or special events (birthday parties, anniversaries, sports tailgating events).

Behavioral information may also be collected to characterize the internal state of the consumer. This includes the time, energy, and financial resources needed to complete tasks in situational contexts and how "depleted" consumers are in these internal states to complete the task. A consumer who is highly depleted in time or energy will tend toward habitual behavior – purchasing on a more emotional (impulsive) basis, than rational basis. Shoppers who suddenly realize their wallet is depleted, may tend to behave more rationally – purchasing on the basis of pricing, or by adhering to more emotional (habitual) behavior by purchasing strictly less expensive store brands without even checking the prices.

Information on internal state of mind is also important when generating behavioral insights. This includes in-context behavioral information about a consumer's mood, sense of control, and stakes/accountability for a decision. A consumer in a stimulated, positive mood may tend to seek new exciting food experiences. A consumer in an agitated, depressed mood may seek to change their state by consuming a comfort food. When the stakes for a decision are high and the choices are many, consumers tend to act fairly rationally in making choice decisions. Such might be the case in making a decision to purchase or prepare at home a special food product for a special event.

Information about environmental and internal states of mind provides the basis for behavioral insights to discover behavioral white space for use or consumption of products in new contexts. Through the discovery of contexts with unfulfilled, yet important jobs-to-be-done, new contexts for use can be established for products. Therefore, it is not so much the focus on context that leads to discovery, but how insights into context segments lead to the identification of unfulfilled jobs-to-be-done by products.

New Habits

The third type of discovery goal is to identify opportunities to change habits. When a product is encountered in a given context by consumers who have specific behavioral tendencies, the underlying source concerns lead to the formation of "surface concerns" and expectations (as defined in Chapter 3). Surface concerns are the expressions of source concerns in a given situational context. Expectations characterize how well a given product might deliver against unfulfilled jobs-to-be-done. These are key inputs into the framework that results in the eliciting of emotions during product experiences.

As discussed in Chapter 3, the basic framework for the formation of emotions depends on surface concerns, expectations, and the stimulus (conceptual or experiential) within a given context. Negative emotions result when expectations are unfulfilled and concerns not met. Positive emotions result when expectations are met or exceeded and concerns met. The types of discrete emotions experienced depend on the type of jobs-to-be-done (i.e. utilitarian, sensory, social, and emotional).

Behavioral insights that lead to the discovery of new habits require more information, as indicated at the third level from the bottom of the Behavior Pyramid. This new information is the characterization of surface concerns and expectations prior to an experience, and an assessment of what emotions are elicited from products within context segments.

Habits are formed through repeated use of products in the same context. These repeat experiences reinforce behavior – giving behavior control over to the unconscious mind. Habits continue due to the formation of emotions generated from cues picked up from the environment and/or from the product itself that motivates use and consumption. Habits are broken when new information is provided that disrupts/awakens the conscious mind.

New habits can only be formed by first breaking existing habits, and then introducing a new experience through an initial trial. The new experience – if it generates positive emotions – forms the basis to establish new cues that motivate the formation of a new habit.

DISCOVERY ITERATIONS

The three types of discovery goals are achieved through a three-step process: opportunity landscape, co-discovery, and opportunity validation. Each step involves one or more iterations of learning. The first step, opportunity landscape, builds pertinent knowledge about consumer behaviors that inspire divergent, and guide convergent thinking in the subsequent steps. Co-discovery is the next step, where ideas are generated for consideration as an opportunity. Opportunity validation is the final step, where multiple ideas are screened, refined or optimized into a defined opportunity solution – achieving discovery goals. Discovery becomes behavior-driven by taking a holistic, iterative approach to discovery learning.

Opportunity Landscape

Before one can generate ideas to diversify thinking about the opportunity in terms of target group, use context, and/or experience, it is essential to have attained sufficient knowledge about pertinent sensing (habitual), seeking, sharing and/or selecting behaviors relevant to discovery goals. Opportunity landscape, within a behavior-driven discovery process, organizes and builds knowledge about these behaviors using the appraisal framework. A number of different methods and techniques can be applied, depending on discovery goals and the base of existing knowledge within the innovation team. This collection of research methods and techniques generate insights into groups of consumer with behaviors associated with unique use contexts and/or experiences that are not fulfilled. They also may involve discovering unfulfilled experiences for target consumers within specific or yet undiscovered use contexts.

These research techniques fall into three categories: knowledge mapping, activity-based research, and social dialogue research. The first helps organize existing knowledge to generate new perspectives for innovation teams, while the other two build new knowledge through research.

Knowledge Mapping

Food companies frequently retain a library of historical research reports. The information within these reports has been earlier referred to as explicit knowledge in the form of consumer insights and consumer product insights.

There is also knowledge of relevant value to discovery that exists as tacit knowledge in the minds of innovation team members (and others currently not on the team). When this information has not extended beyond its shelf life, it can often be integrated into new insights through the use of the appraisal framework.

In each of these cases, opportunity assessment can be envisioned by means of a schematic called a brandTrace (see Figure 7.2). The grid seen upper left is the "Behavior Grid" – occasions (i.e. use contexts) divided according to surface concerns that are frequent, important, and unfulfilled. The appraisal framework (Emotions Insight Wheel, Chapter 4) is represented upper right. Sufficient information about the behavior drivers typically exists to enable inferences to be drawn to map each behavior driver to an emotion. In addition, a product grid (lower right) can be generated to relate sensory attributes to product cues. These cues can take on different dimensions, such as type of cue, and whether the cues came from the messaging, packaging or the package contexts. These cues can also be mapped into specific emotion. The behavior grid can also take on different dimensions such as target consumer, occasions, and types of behavior drivers (functional, sensory, social, psychological) with specific behavior drivers mapping to specific emotions.

This approach tends to generate heuristic models where "gaps" (unknowns) can easily be filled by "inferences." For example, a food company was seeking to discover what might be the best new experience that fit with a novel technology application. They had identified a use context for a product oriented at

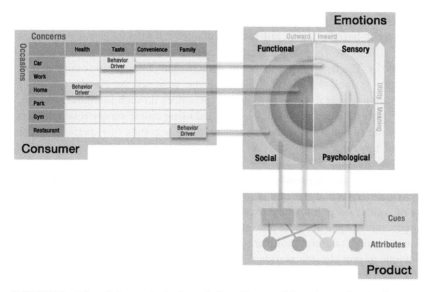

FIGURE 7.2 A knowledge map in the form of a brandTrace used for understanding the discovery landscape, with behavior drivers and associated cues mapped into the Emotions Insight Wheel. (Please refer to color plate section)

a specific target group of kids. All of their recent historical research involved qualitative research (i.e. ethnography and focus groups). Much of this recent information was readily mapped into the brandTrace structure. Whereas the qualitative research did not explicitly focus on product emotions, the quality of the information on cues and behavior drivers enabled a brandTrace to be generated with emotions inferences. This provided the innovation team with a clear understanding for what combinations of drivers and cues might be used to generate a wide range of new ideas and a new experience.

This new knowledge (i.e. repackaged information into insights) from research such as this can sometimes avoid the cost and time involved in repeating past research. However, in other situations, little might be known about the occasion (context of use), or the importance, fulfillment, or frequency of behavior drivers (surface concerns) within those occasions. In these cases, new research must be conducted to build up a base of knowledge sufficient to inspire divergent thinking.

Activity-Based Research

Activity-based research involves generating information into the landscape of behaviors within which to discover new target groups exhibiting unfulfilled source concerns in various contexts. It can also be applied to simply increase knowledge into how target groups are using products in different use contexts. Whereas more traditional research tends to be more question-based, this type of research is more activity-based – collecting information about consumer activities that roll up to define behaviors.

Activity-based research utilizes a number of different tools and techniques. These all involve engaging consumers as research participants and assigning them tasks. The tasks they conduct provide insights into specific activities that characterize their behaviors. Tasks can be assigned to simply providing information to researchers about their experiences through journals or diaries. This type of research is typically conducted using online survey research tools or specific websites designed for self-reported accounts of experiences. The form of this information tends to be quantitative and qualitative (open-ended text) information.

However, this approach to research can be extended to capture more complex activities. By assigning more complex activities, research can become more "immersive." Immersive research techniques combine four technologies to generate deeper insights: the Internet to provide a channel for communication, social networking software to provide peer-to-peer engagement, web cameras to provide information richness about context, and mobile devices to provide research "in-moment experience" immediacy.

The value of activity-based research is to deepen insights by listening to consumer stories about their experiences. These stories tend to follow the natural way that human beings tend to communicate their experiences – describing the setting (i.e. what was the context), actors (i.e. who were people associated with the experience), what was to be done (i.e. what were the concerns to be fulfilled or jobs to be accomplished), and what were the outcomes (i.e. what happened and how did the outcome make one feel).

It is in these stories that rich information emerges uncovering insights of value to the discovery process. The key to the design of behavior research is to structure the information generated from these tasks into the appraisal framework, so as to be able to integrate the hybrid of quantitative and qualitative information into insights. Text information can be easily converted into quantitative categorical information. More complex qualitative information, such as pictures and videos, must be "tagged" into specific categories of information. The appraisal framework provides an ideal basis to structure tags and qualitative conversions into quantitative content.

In 2009, InsightsNow conducted research to understand the landscape of occasions with the greatest opportunities for new healthy food products for a specific target group of consumers.[2] The appraisal framework was deployed to guide how information was collected, categorized, and integrated into insights. Occasions were categorized as ready-to-eat snacks, home-prepared snacks, home meals, and meals outside of the home.

To complete this research, 600 participants were tasked to post their food consumption experiences into an online diary. Over a 3-week period, participants posted over 5,000 occasions, including where, when, and with whom food was consumed, as well as what "jobs" were to be done by the food (product driver). Participants also logged information about the foods and brand(s) consumed, as well as the importance of each job and how well it was fulfilled by the product.

An adaptive, iterative research design was employed where the questionnaire quickly evolved from 100% open-ended to describe the jobs-to-be-done, to check-all-that-apply (with always a few additional open ends). Finally, participants were asked about what barriers (if any) were stopping them from not getting the experience they really sought. Additional information was collected about consumers, such as their demographics and source concerns (see Figure 7.3).

In this research, it was found that the biggest behavioral differences were among consumers with a high proportion of occasions of eating alone (see Figure 7.4). This group was characterized as eating 80% of all dinner home meal occasions alone – even though 50% were married. Using a measure of market "reach" called a Shapley Value,[3] it was found that in the context of

---

2. D. Lundahl, J. Schaefer, G. Stucky, S. Bryles and J. Bryles, Getting the Health and Wellness Strategy Right. Presentation at IFT Wellness 09: At the Forefront of Food and Health, March 26, 2009

3. W.M. Conklin and S. Lipovetsky, Marketing Decision Analysis by TURF and Shapley Value, International Journal of Information Technology & Decision Making, 4(1) (2005) 5–19

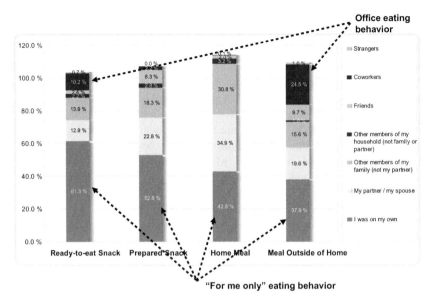

FIGURE 7.3 Results of activity-based research into food consumption behaviors for four types of foods (ready-to-eat snacks, prepared foods, home meals, and meals eaten outside the home). (Please refer to color plate section)

a dinner meal at home the most important behavior drivers were "satisfy your hunger" and "taste good." However, these did not distinguish between these "eating alone" and "sharing a dinner at home" experiences. The most important "eating alone" behavioral drivers (surface concerns) were that the meal be "quick," "convenient," and "keeps you healthy." The most important "sharing

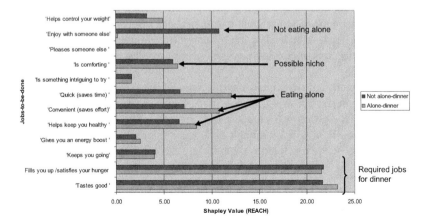

FIGURE 7.4 Differences in importance for various jobs-to-be-done by "at home" meals when eating alone or with someone else. (Please refer to color plate section)

a meal" behavior drivers were "enjoy with someone else" and "please someone else." Further, it was found that "is comforting" scored moderately high for both groups. In addition, it was discovered that "is something intriguing to try" was the most important, yet unfulfilled driver. In response to the question about "what is stopping you from getting what you want," a surprisingly large number of dinner home meal occasions included "not enough time," "not enough money," "providing for someone else (not for me)," and "too tired to prepare."

The diary research study applied the appraisal framework to structure how information was collected and integrated into insights. These insights provided the basis for a story about what consumers – eating a dinner alone or sharing a dinner at home – are seeking. The application of the appraisal framework (Emotion Insight Wheel™) leads to new inferences and insight for how these behavior drivers lead to satisfaction, dissatisfaction, enjoyment, pride, and intrigue. This story provides a strong basis for inspiring innovators to creatively generate ideas as to how to deliver product experiences against these behavior drivers, in different contexts leading to specific emotional impact.

The extension of activity-based research to provide visual media (pictures and video clips) about in-moment, in-context experiences adds an even greater richness of information to integrate into insights. The challenge in capturing richer qualitative content such as this, is in how to convert this information into insights. However, the appraisal framework provides an outstanding structure to guide how this information might be converted into quantitative information, to be integrated into insights using statistical models or might be directly integrated into insights using heuristic models.

Social Dialogue Research

A third technique of value in understanding the opportunity landscape is to sample the river of information of peer-to-peer dialogue to understand emotional impact about consumer experiences. In Chapter 6, a technique was introduced to track brand attitudes over time through sentiment analysis. This technique can also be applied to gain insight into the emotional impact of product experiences within specific categories.

Consider the example of sentiment captured about differences between two well-known hamburgers – the Big Mac (McDonald's Corporation) and the Whopper (Burger King).[4] In this case, sampled dialogue about how the bread or the meat impacted the experience with either of these two products was captured (see Figure 7.5). By applying sentiment analysis, the directionality of the emotions and the projection of the emotion (the bread or meat) was used to gain key insights into these two product differences. In addition to these

---

4. T. Ting and A. Pettit, Battle of the Burgers: Using Social Media Research to Deconstruct Satisfaction with the Big Mac and the Whopper, Unpublished data. Provided by permission of Conversition, Inc., 2011

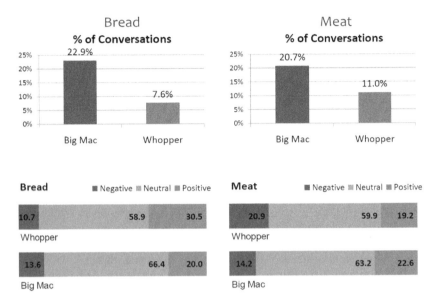

FIGURE 7.5 Sentiment analysis provided by Conversation from online social networks with regard to the bread and meat from Whopper (Burger King) and Big Mac (McDonald's Corporation). (Please refer to color plate section)

sentiments, the actual words used by consumers describing their emotions were also analyzed. This provided additional insight into the language of emotions being used by consumers in the context of their burger eating experiences.

The value of knowledge gained through the opportunity landscape step cannot be overstated. This organization of existing knowledge and creating of new knowledge, through activity-based and social dialogue research, provides the spark that innovation teams often need to creatively generate ideas for what are the possibilities to take consumers to the destinations that they seek.

OPPORTUNITY DEVELOPMENT

Once the opportunity landscape is known, the second step in the discovery process is to apply opportunity landscape knowledge, broadening the range of ideas for how to achieve discovery goals. This is called opportunity development. It involves the creative process of divergent thinking to discover a range of possible destinations for target consumers.

Whether the goal is to discover new consumer targets, new contexts of use, or new ways to change habits, the generation of ideas always involves consideration about the experience. This is accomplished by combining the behavior drivers identified from the discovery landscape into themes that conceptually define potential experiences. Opportunity development involves building and scoring themes to ensure that they have the potential to deliver emotional impact.

Theme Building

Theme building involves an iterative process of creative generation of ideas for potential themes. Each theme is comprised of behavior drivers, potential cues, inferred emotions, a positioning statement, and a name.

An approach to theme building, that has proven helpful, is to add some structure to the process. This process does not hinder creativity, but focuses it to characterize some aspect of the opportunity landscape. The following seven steps are used to build themes:

1. **Build behavior drivers list:** Generate a list of known, or hypothesized surface concerns, believed to drive behaviors through existing product experiences within use contexts for various consumer groups.
2. **Build cues garden:** Generate lists of cues that are known, or hypothesized, to elicit emotions from existing product experiences within use contexts for various consumer groups.
3. **Add inferred emotions:** Map behavior drivers (surface concerns) to emotions and cues to emotions using the emotions insight wheel. Note: this is typically done concurrently within step 1 and 2.
4. **Build themes:** Generate "prospective opportunity solutions" by creatively identifying groups of behavior drivers that can be associated into themes.
5. **Link cues to themes:** Identify cues that align with each theme and which have the same inferred emotions as the respective theme, behavior drivers.
6. **Build position statement:** Creatively generate the simplest statement that communicates the theme. Note that simplicity can often be gained by not duplicating behavior drivers that can be built into an opportunity solution unconsciously through a theme's respective cues.
7. **Name theme:** Creatively develop name for each theme from their respective positioning statement.

In addition, one or more benchmark themes can be created from existing benchmark products or product lines in the marketplace. These benchmarks themes are "re-engineered" from the known product positioning statements and the same collection of behavior drivers, cues, and inferred emotions.

Theme Scorecards

The final step in opportunity development is to score themes (prospective opportunity solutions and benchmarks). Theme scoring provides a way to gauge the effectiveness of the innovation team, in order to discover alternative opportunities. The following three metrics can be used to score themes:

1. **Behavior Differentiation** – the number of behavior drivers used in a theme.
2. **Emotional Reach** – the number of emotions inferred by the theme.
3. **Creative Potential** – the number of cues conceived of as potentially associated with the theme.

These three metrics gauge the potential for an opportunity, not in terms of a market volumetric forecast, but in terms of behavioral impact. At the "focused" front end of the product development process, it makes no sense to forecast market volume for specific themes. It does make sense to estimate total market size (market volume), within which an opportunity solution targets. However, this should have been done as part of the market analysis in establishing brand strategy.

These three metrics gauge a theme's potential to be developed into one or more products that will achieve "behavioral impact" throughout a target market. High behavior differentiation implies a product developed from a theme has high potential to be perceived as differentiated in the marketplace by target consumers within given contexts of use. High emotional reach implies a product developed from a theme has high potential to elicit different types of emotional impact throughout the target market. High creative potential implies the bigness of the theme to be developed into a line of different products that target the same market opportunity.

Each of the themes and their respective scores can be displayed within a scorecard. A scorecard provides a simple way for themes to be easily assessed, compared and (if needed) refined. A scorecard can be read as in Figure 7.6.

Consider again the Instant Specialty Coffee Case Study with "Nora." Recall that she had a positive trial experience with the instant specialty coffee.

How to Read a Theme Score Card

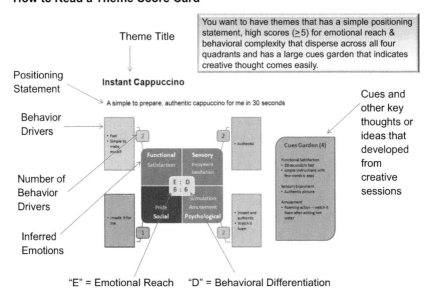

"E" = Emotional Reach   "D" = Behavioral Differentiation

FIGURE 7.6 A scorecard for themes generated to characterize potential consumer product experiences. (Please refer to color plate section)

This experience included feeling functional satisfaction that the product was fast and simple to make, sensory satisfaction and enjoyment in the authenticity of the consumption experience, intrigue (stimulation) from the picture on the package front panel, and amusement in watching the foam grow after adding hot water. In addition, it was learned from Nora that there are possible behavior drivers, such as pride, that she can make a specialty coffee experience just for her, as well as for her girlfriends, that often come over to her home.

This research information can be restructured into behavior drivers, cues and associated inferred emotions – the building blocks of themes. From these building blocks, two themes were generated: "Instant Cappuccino" and "Secret Barista." These themes were built into scorecards, as defined above. "Instant Cappuccino" theme scores a 6 for Behavior Differentiation, 6 for Emotional Reach, and a 4 for Creative Potential (Figure 7.7). The "Secret Barista" theme scores higher, with a 7, 8, and 5 for the same three respective metrics (Figure 7.8).

Through these seven steps, a process exists to help innovation teams apply the knowledge gained in opportunity landscape. The process is designed to inspire creativity focused on achieving Discovery-phase goals. The metrics built into scorecards provide metrics to gauge the effectiveness of this process. For example, if the "Instant Cappuccino" theme were a benchmark, a scoring

Instant Cappuccino

A simple to prepare, authentic cappuccino for me in 30 seconds

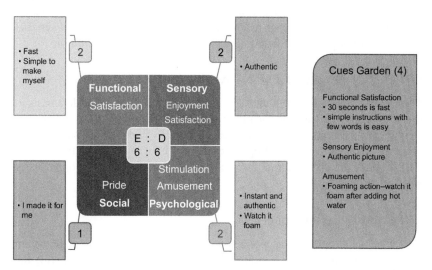

FIGURE 7.7 The "Instant Cappuccino" theme developed for the "Nora" archetype. (Please refer to color plate section)

Secret Barista

My easy, fast, secret way to make an authentic cappuccino for friends

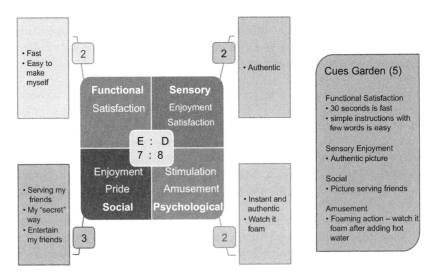

FIGURE 7.8 The "Secret Barista theme" developed for the "Nora" archetype. (Please refer to color plate section)

system exists for innovation teams to gauge their creative effectiveness in generative alternative themes that have the potential to be differentiated from the benchmark functionally, sensorial, socially, and/or psychologically; generate more emotional reach by eliciting emotional impact in more ways; and provide more degrees of freedom to build cues into a product solution.

OPPORTUNITY REFINEMENT AND VALIDATION

The creative generation of themes and respective scorecards provide a basis to design research that extends learning to achieve discovery goals. There are a number of research techniques that can be used to guide the decision making process to screen, refine, and optimize themes into an opportunity solution. Further, there are also different research techniques to validate the opportunity solution through consumers.

Opportunity Refinement

The objective of opportunity refinement is to refine the collection of generated themes into an opportunity solution. This is best accomplished by applying the iterative approach to emotions research, as discussed in Chapter 5. Iterative research gains efficiency in learning by exposing target consumers, as research participants, to themes in the form of product or product line concepts and engaging consumers in various forms of dialogue to assess their relative emotional impact. Learning occurs when the innovation team is able to use insights from this research to guide decision making about the optimal solution. This learning cycle can be accelerated by applying the principles of adaptive learning.

A number of research techniques exist to generate research information from the appraisal of concepts by research participants. Which technique is most appropriate depends on the number of generated themes and how distinct the themes are from each other. When there are very few themes and they are each distinct from one another – a simple forced choice consumer testing design, with appropriate statistical analyses (multinomial, non-parametric, or analysis of variance), can prove effective. Choice is recommended over ratings-based techniques, due to the lack of sufficient context to warrant purchase intent, purchase interest, or liking ratings.

When the number of distinctly different themes grows beyond what might be easily appraised by research participants, then an interim choice-based screening or filtering technique might be deployed to reduce the number of themes to a more manageable number. Alternatively, choice-based quantitative techniques such as tournaments, ranking, and max-diff might be applied to screen and test in one study. These more classical concept testing techniques will be covered in more detail in Chapter 8.

The more typical situation will have a large number of less distinct themes for refinement into an opportunity solution. While a choice-based approach remains optimal, the fact that themes are less distinct suggests the need to apply a choice-based conjoint modeling technique.

Choice-based conjoint[5] has been applied widely to understand the relative importance of elements of packaging or concepts to consumer impact. While the concept and/or packaging is assessed as a whole, there is an underlying model that characterizes which elements contribute to overall impact. This modeling approach fits well with the application of the appraisal framework to build themes. The content on the theme scorecards can be used to develop an underlying conjoint model with behavior drivers (surface concerns) and cues as model elements.

For example, consider the two themes generated above from the Specialty Coffee Case Study. The conjoint model (see Figure 7.9) includes 10 behavior drivers and six cues. Through research design, where all participants appraise each theme, separate model coefficients can be estimated for each consumer. This allows for the estimation of importance factors for the whole sample of consumers, or breakout groups representing any segment of the target market.

5. K. Chrzan and B. Orme, An Overview and Comparison of Design Strategies for Choice-based Conjoint Analysis. http://citeseerx.ist.psu.edu/viewdoc/download?doi=10.1.1.87.597&rep=rep1&type=pdf, 2000

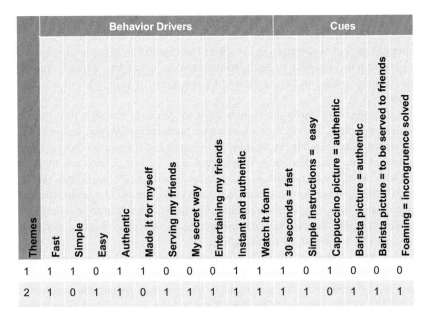

Themes		Behavior Drivers										Cues				
	Fast	Simple	Easy	Authentic	Made it for myself	Serving my friends	My secret way	Entertaining my friends	Instant and authentic	Watch it foam	30 seconds = fast	Simple instructions = easy	Cappuccino picture = authentic	Barista picture = authentic	Barista picture = to be served to friends	Foaming = incongruence solved
1	1	1	0	1	1	0	0	0	1	1	1	0	1	0	0	0
2	1	0	1	1	0	1	1	1	1	1	1	1	0	1	1	1

FIGURE 7.9 The explanatory variables for the first two themes (Instant Cappuccino and Secret Barista) used to build a choice-based conjoint model.

The estimation of which behavior drivers and cues are significant importance factors provides insight into new combinations for an optimal opportunity solution. This opens up a number of optimization methods where a new – untested – theme can be identified for individuals, breakout groups among the sample of consumers, or for all research participants.

Opportunity Validation

The purpose of validation is to ensure that the discovery goals have been achieved. However, a separate validation study is not always necessary. When a benchmark exists and themes are distinct (i.e. few overlapping behavior drivers and cues), it is possible to build validation right into the refinement step by establishing statistical criteria, e.g. the test theme is significantly chosen over the benchmark. When themes are less distinct and the refinement process has led to the identification of an optimal theme, then the presence of themes provides for one of two validation options. The first is to build into the refinement step a test where the predicted theme response is tested statistically against the benchmark. The alternative is to set up a final validation test where the optimized theme is tested head to head against the benchmark.

The biggest challenge exists when there is no benchmark. Such may be the case when the goal is to discover a new use context for an existing product. In this case, however, it is possible to establish internal statistical criteria into the refinement study as validation. One approach is to build a database of successful themes and their actual number of significant behavior drivers and cues from conjoint models. These can provide excellent metrics when they have been proven to track with product success in the marketplace.

ACCELERATING THE BEHAVIOR-DRIVEN INNOVATION PROCESS

It is through this focusing at the front end of the innovation process (the discovery process) that the whole innovation process is accelerated. Likewise, in identifying clear opportunity solutions based upon behavioral criteria, the subsequent phases of the innovation process will achieve greater focus.

The focus gained by identifying chocolate gift-givers as a new target segment is significantly less than the focus gained by the discovery of the contexts whereby M&Ms might serve to achieve specific jobs-to-be-done. The further insight that personalization is a missing element of experience serves to further focus the discovery process on the next phase of the innovation process – the scoping of ideas for how to fulfill the given jobs through the sale of personalized M&Ms to chocolate gift-giving consumers.

Key Points

- Behavior-driven discovery is a highly focused front end to the product development process. This focus comes through strategy development into where to play, how to play to win, what to innovate, and the type of innovation to undertake.
- The product development process includes four phases (Discovery, Define, Design and Development). Each phase involves a more refined, granular understanding of consumer behavior within tighter guardrails.
- The goal for discovery is to identify the biggest opportunity to impact the behaviors of target consumers. This is achieved by discovering product experience themes that engage new consumers, motivate new ways to use or consume products, and/or form new habits.
- The process of discovery involves understanding the opportunity landscape and creatively developing product experience themes within that landscape. This is followed by the refinement of themes into an opportunity that can be validated.
- Knowledge mapping is a method for building the discovery landscape by placing existing explicit or tacit knowledge into the appraisal framework. This framework results in new knowledge about the relevant concerns that exist within situational contexts of interest and the cues that exist within packaging and packaged foods, and how these concerns and cues map into the Emotions Insight Wheel.

- Activity-based research involves assigning tasks for research participants to capture information about their behaviors and the underlying factors motivating those behaviors.
- Social dialogue research involves sampling from the river of information generated by peer-to-peer dialogue, through online communities and social networks. This information provides insights into the emotional sentiment of different product experiences that motivate behaviors.
- Theme building is the creative process for generating ideas based upon the discovery landscape into potential new product experiences – i.e. destinations that target consumers seek.
- Scorecards facilitate theme building by focusing theme development on the development of simple positioning statements that lead to high behavioral differentiation, high emotional reach and high creative potential.
- Opportunity refinement involves consumer research into the emotional impact of themes. Choice-based conjoint can lead to the discovery of themes within guardrails that achieve high emotional impact reaching across the target consumer.
- The opportunity solution is a refined theme for a new product experience that has been validated. Validation can be achieved through testing themes against benchmark themes with known behavioral outcomes. In the absence of a benchmark, internal validation can be achieved by measuring the actual behavioral differentiation and/or emotional reach as action standards.

Chapter 8

Define

Intuition and concepts constitute ... the elements of all our knowledge.

Immanuel Kant

CONCEPT DESIGN

"Define" is the first of two design phases required to develop a successful product. Both phases are focused on design – the process of building cues into products to deliver emotional impact. The first involves the design of the product concept. The second involves the design of the tangible aspects of the product to be built.

Those familiar with the Stage-Gate process may consider the define phase as stage 1 – the Scoping stage (see Cooper[1]). Product scope typically refers to the "features and functions that characterize a product."[2] In the case of a behavior-driven innovation process, the concept is the heart and soul of a product. This definition goes beyond the traditional definition of a product concept. It defines the concept as a collection of cues fitting under a common theme that elicits emotional impact.

For this reason, innovation team members use the concept as a target for subsequent design and development. Marketers communicate the concept through various channels such as television or digital media advertisements. Package designers develop the package design to express the concept. Food designers use it to guide their food creations. This commonality among all elements of the product leads to harmony.

Harmony is essential for the consumer to come to know a product. The concept is the cognitive structure (i.e. schema[3]), providing a basis for consumers to experience products. When marketing messaging, package design, and the design of the packaged food are in harmony with the concept, it is much easier for consumers to become habitual users of products. Disharmony leads to confusion and the awakening of the conscious mind to reconcile

---

1. http://www.prod-dev.com/stage-gate.php. January 10, 2011

2. http://en.wikipedia.org/wiki/Scope_%28project_management%29, February 25, 2011

3. http://en.wikipedia.org/wiki/Schema_%28psychology%29, February 25, 2011

---

differences between what is promised and delivered through a product experience.

The Define phase follows naturally from the Discovery phase. In the previous chapter, we defined a process for innovation teams to discover the opportunity solution. This solution is the theme under which all product concepts must fit for a given product line. The cues identified in the Discovery phase become the guardrails from which to build concepts. Each product in the line must deliver an experience that fits with this theme, yet do so in different ways to achieve emotional reach across the consumer target.

In many ways conceptual design is similar to discovery. The knowledge generated to this point becomes the basis for new learning. In the early iterations of conceptual development, knowledge is applied to inspire creative design of a number of candidate concepts for each product within the line. Knowledge is then used to subsequently guide concept refinement and validation.

In this chapter, research methods and techniques will be introduced to generate consumer insights that identify consumer language associated with product experiences. These insights will focus the innovation team to develop concepts within guardrails that not only motivate trial, but also provide opportunity for trial, by simultaneously disrupting the unconscious mind and breaking habitual use of incumbent products.

CONCEPT APPRAISAL

The appraisal framework first introduced in Chapter 4 provides the basis for concept design. The emotions associated with the appraisal of concepts are different from those of product experiences. The behavioral drivers associated with the opportunity narrow the scope of what possible cues can be built into the concept. The usage cases for the product focus the conceptual design on the specific jobs-to-be-done. In this section, the purpose of conceptual appraisal will be discussed, not only to help define the product conceptually for product development, but to ensure that the product concept will motivate important behavioral changes for the target consumer.

Maslow's Hierarchy

The marketer has traditionally owned concept development. Marketers – in their role – have historically approached marketing by engaging the conscious mind of consumers to increase awareness and positioning products to appeal to, or be concerned with, fulfilling need states. This approach to marketing goes back to the 1940s. In 1943, Dr. Abraham Maslow published his landmark article "A Theory of Human Motivation."[4] In this

---

4. A.H. Maslow, A Theory of Human Motivation, Psych. Review, 50 (4) (1943) 370–396

article, he proposed a framework for understanding human motivation using a needs-based hierarchy. The ideas put forth in this article were later expanded in the first edition (1954) of Maslow's book *Motivation and Personality* (1987).[5] In this book, he formally introduced the "Hierarchy of Needs" that has been a foundation for marketing for over 50 years. His "Hierarchy of Needs" includes six levels of increasingly complex needs (physiological, survival, safety, love, esteem, self-actualization). The lowest needs are the most basic that humans will trade off in importance when faced with a choice. The highest is the need to find fulfillment and meaning in life.

Maslow's framework has shaped many of the methods used by marketers today – applying a rational-based approach to market products. However, the emotions-based appraisal framework replaces Maslow's Hierarchy with more contemporary scientific evidence that it is emotions, not need states, that motivate most human behavior. Yet, there are fundamental truths to Maslow's Hierarchy. Some need states are more important than others. Just as some surface concerns are more important than others at the moment of an appraisal. The appraisal framework thus leads to the conclusion that some discrete anticipation emotions associated with conceptual appraisal are also more motivating than others. They achieve greater emotional impact.

This logic leads to an opportunity to develop a hierarchy of anticipation emotions for application in concept development. Consider again the four quadrants of the Emotions Insight Wheel (Chapter 4). Can the four quadrants and their respective concerns be associated with some or all of Maslow's need states? It appears that this can be done. However, surface concerns with their more fundamental source concerns provide for a broader set of behavioral drivers than need states.

Consider the schematic in Figure 8.1. Functional concerns tend to include those associated with Maslow's lowest, most basic needs. This includes physiological needs – i.e. states of mind rooted in needs to fulfill hunger, thirst, sexual drive, etc. For example, hunger as a need state may become a behavior driver through a heightened concern to get filled up. However, it is the emotion – of anticipation emotion of desire – that actually motivates action, not the need state. Similar arguments can be made for most need states defined by Maslow.

The next two levels in his hierarchy are the need for survival and safety. The need for being safe may lead to a heightened concern for not getting sick in a situation where a particular food is being appraised with a smell or look that cues "not safe." However, it is the emotion of fear, not the need state to be safe that motivates avoidance of the food. Maslow's need states do not fully address

---

5. A. Maslow, Motivation and Personality, third ed., HarperCollins, New York, NY, 1987, pp. 293

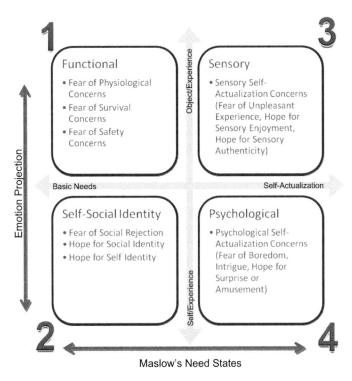

FIGURE 8.1 Relationship between Maslow's need states and the surface concerns and emotions of the appraisal framework.

the question of how cues in the environment, or from within food products, may motivate behavior.

The needs from the fourth and fifth levels of Maslow's Hierarchy can be associated with concerns with the Self-Social Identity quadrant. This includes associating needs to concerns over social-identity (the need to belong to the herd) and self-identity (the need for self-esteem). For example, the need to be a good mother may lead to a heightened concern at the grocery store to seek out a more healthy food that her child will accept. However, it is the emotion of hope that a more acceptable healthy food can be found, rather than her need for self-identity (to be viewed by herself as a good mother) that motivates her actions to seek out a solution. The behavior driver leading to hope is her immediate concern within the shopping experience to find a more healthy food that her child will accept.

Sensory and psychological concerns are more easily associated to the uppermost needs in Maslow's Hierarchy – i.e. self-actualization (i.e. to find fulfillment, purpose, and meaning to one's life). They include sensory concerns to avoid unpleasant experiences and to aspire to food experiences that are enjoyable and, at times, authentic. They also include psychological concerns that aspire to combat boredom through experiences that are surprising, intriguing, and amusing. For example, the needs for self-actualization might lead to a heightened concern to select a food for dinner that is comforting after a difficult day at work. However, it is the hope that a comfort food will be found in the food pantry that actually motivates the selection of a product for dinner, not the need state for self-actualization. The hierarchy of needs provides a general way to rank concerns and their respective emotions in striving to get to desire – or in some cases to elicit disgust for competitive products. However, needs tell only a part of the story as to why consumers are motivated to engage in sensing, seeking, selecting and sharing behaviors. Need states do not address how cues built into product concepts or products lead to anticipation emotions.

As discussed in Chapter 4, the anticipation emotions associated with conceptual appraisal are fear, hope, intrigue, desire, and disgust. From the Emotions Insight Wheel, fear is more prevalent in the functional quadrant with regard to more basic needs. While there is some fear associated with social rejection, most of the self-social identity concerns lead to hope. Sensory concerns also tend to mostly involve hope. However, it is conceivable that some consumers may experiences some fear of unpleasant sensory experiences. Psychological concerns tend to lead to hope or intrigue. These three anticipation emotions (fear, hope, and intrigue) provide the basis for desire and disgust. Desire is an overarching positive emotion motivating choice behavior (i.e. action readiness to possess). Disgust is action readiness to distance oneself from a product. Motivation is driven predominantly by emotions, not need states.

Hierarchy of Anticipation Emotional Impact

Marketers, and others on the innovation team that apply a framework that goes beyond Maslow's needs state, have an advantage. However, Maslow's arrangement of need states as a hierarchy does lead to a question: Is there a hierarchy of concerns that might be applied to drive the conceptual design of products?

The associating of concerns and anticipation emotions. With Maslow's Hierarchy of Needs, leads naturally to a generalized ranking of concerns with regard to their potential for emotional impact. This ranking is as follows: functional fears>social rejection fear>self-social identity hopes>sensory fears>sensory hopes>psychological fears>psychological intrigue>psychological hope. However, these rankings should be followed cautiously. The appraisal framework is clear that culture, personality, values, and other identifying qualities about consumers underlie the relative importance of which concerns lead to greater emotional impact. For example, a consumer with a personality to be more preventative (than promotional) may tend to have stronger behavior drivers through fear (than hope). However, this ranking provides the basis for conceptual design that is more than simply an interim step in the product development process. It becomes a key tactic to change consumer behavior.

Disrupting and Reinforcing Habits

In Chapter 5, it was argued that in order for a strategy to take market share away from a competitor, the brand must be able to move the consumer through a series of four appraisal moments. This sequence includes appraisal moments that disrupt old habits (i.e. using the incumbent product), gain trial by introducing an alternative, generate a positive experience from the alternative experience, and reinforce the alternative experience to establish a new habit with the alternative (see Figure 8.2).

Also in Chapter 5, the "4 S's of consumer behavior" were introduced (Sensory, Seeking, Sharing, and Selecting). It is worthy to note that conceptual appraisal is essential to disrupting old habits and getting trial for an alternative. Typically, innovation teams focus on the market message and cues associated with getting trial. However, trial is difficult – unless there is first disruption to disengage the unconscious mind and awaken the conscious mind. Therefore, the most effective strategies in concept development are both disruptive to break habits and engaging to get to trial.

This disrupting and engaging may not only involve sensorial and selecting behaviors. It may well be that disrupting and/or engaging may come from seeking behaviors (e.g. an Internet search about a topic concern or solution) or sharing behavior (e.g. peer-to-peer sharing about a new concern or experience). These links between seeking, sharing, and selecting suggest that the marketing messaging must be closely integrated into the conceptual design of a product. This linking of marketing and innovation tactics can take advantage of the weaknesses of a competitor (i.e. a concern that can lead to fear). Further, this linking enables the strength of a new product (e.g. aspirations to

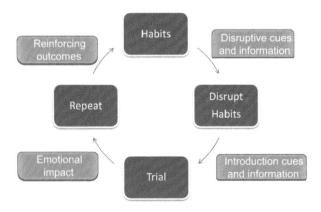

FIGURE 8.2 A behavioral strategy for disruptive innovation.

achieve a self-actualizing concern) to become known as a competitor's weakness.

Research 2.0

The emergence of the prosumer as discussed in Chapter 1 and again in Chapter 5, changes the way in which the "4 S's of Consumer Behavior" contribute to market success for newly developed food products. The tactics used in the conceptual design of products must factor in these relative strengths and weaknesses for developed products against competitors. While selecting behavior is essential for trial, it is sharing and seeking behaviors that will drive market dispersion. It is sensorial behavior that leads to the establishment of new habits with the newly developed product.

This new knowledge on how consumers behave must be understood in order for products to be accurately and successfully developed and marketed. The researcher's role is to ensure this behavioral knowledge is woven into tactics throughout the scoping phase of product development.

Ray Poynter,[6] in a Research World article, calls for a paradigm shift in how researchers respond to not just the prosumer, but also to the way technology is enabling a new research methodology. Poynter suggested moving to "Research 2.0," a new research model that is more participatory. This approach to research is less scientific, but more qualitative in that it is design oriented, non-statistical, collaborative, and advocates co-creation with consumers. These thoughts align with the "Learning Cake" and rapid iterative learning cycles discussed in Chapter 5.

Although Ray Poynter's perspective is evolving, the behavior-driven approach to innovation calls for an extension of Research 2.0 – to apply the appraisal framework to not just measure the changing concerns and behaviors of prosumers, but to deliver products that enable the behavioral changes they seek. This approach brings into the research model more than just the psychology of emotions as a framework to integrate quantitative and qualitative research into insights. This approach brings new thinking into research about how to enable consumers to achieve the behavior changes they seek.

DEFINE GOALS

The goals for the scoping phase fall naturally from the behavior plan implicit within the innovation strategy. A strategy to apply incremental, emergent innovation typically requires simply reinforcing existing habitual behaviors. For example, a strategy of incremental line extension is sufficient to keep pace with the changing preferences of consumers. In this case, scoping phase goals

---

6. R. Poynter, Can Researchers Cede Control? What Web 2.0 Means to Market Research, Research World, July 2007

should reflect the strategy to reinforce, not change behaviors. A strategy to apply disruptive innovation requires disrupting existing habits, engaging the consumer in trial, and reinforcing new habits. For this, scoping goals must both disrupt the use of the incumbent and position the new product for initial trial.

The underlying strategic focus (type of platform) for innovation should also be reflected in the scoping goals. If the innovation project is technology-driven, then the competitive advantage of the technology should be used to position each product against the competition. Likewise, if the innovation is market driven, then the goals for the define phase should be to ensure that the unfulfilled concerns identified in the market are met.

Guardrails

How these goals are to be achieved depends on the guardrails set up from the discovery phase. From a concept design perspective, the innovation team has considerable creative leeway in how the collection of cues identified in discovery can be built into product concepts. Cues can come from the product name, words used in positioning products, imagery, and elements that exist within the environment in which a product is conceptually appraised. Not all cues envisioned from the discovery phase need be built into each product. Furthermore, different segments of target consumers may not respond to these potential cues in the same way.

The refined theme from the opportunity solution in Discovery is the starting point for concept development. This theme includes the qualities for the line that attract seeking behavior and motivate selecting behavior (trial). These qualities must portray the brand personality, and drive behavior through a position statement for the line, imagery, and additional key words and phrases that cue emotional impact.

THE DEFINE PROCESS

Conceptual design is a highly iterative process. In the early iterations, there needed to be sufficient knowledge to inspire creativity and generate ideas. The knowledge for creative development of concepts comes from the concept landscape. This concept landscape is more granular than the landscape used for refining the opportunity solution. It not only includes the behavior drivers and cues, but also the language and imagery of the consumer in communicating and understanding product concepts.

This concept landscape is also defined by its guardrails. These guardrails are different from those set up at the front end of discovery. They focus concept development on a more narrow set of ideas that each promises a differentiating experience driving behavior change. This concept landscape is the knowledge base within which to creatively build into concepts, the behavior drivers, and cues that provide meaning to the product and lead to emotional impact.

Once this landscape has been built, a basis exists for innovation teams to move to the second step of Define – concept development. Concept development starts with the building of a concept template – that includes all behavior drivers and cues that will be fixed for the whole product line. Concept development involves generating and filtering ideas that fill out this template, to generate specific product concepts for each product in the line.

The generation of ideas tends to be a highly creative, iterative process. The approach to creative development involves collaboration between different members of the innovation team – typically innovators and researchers, but more frequently partners and consumers. These ideas must then be filtered and often combined into a set of product concepts that define the line.

The line concepts must sufficiently cover the concept landscape to achieve reach. Once concepts have been developed, they often must be refined or optimized into final concepts for validation. In these iterations, knowledge is gained to make decisions about the conceptual design for each product in the product line. These iterations often involve screening through a large number of concepts, to filter out concepts with lower emotional impact, and to identify the combination with greatest reach in three ways – behavior drivers, cues, and emotions. This process of refining can also involve optimization where the best mix of behavior drivers and cues leads to greatest emotional impact.

This iterative process of scoping landscape, concept development and refinement leads to a final validation. As with the Discovery phase, validation may be built into the final iterations of concept refinement or be achieved through its own iteration of learning. Validation may involve benchmark concepts or decision metrics, based upon historical knowledge for what behavioral measures signal success in achieving Define phase goals.

THE CONCEPT LANDSCAPE

The language of the selecting or seeking consumer involves more than simply purchase interest. It involves a description of the destinations consumers seek and would be motivated to select. The concept landscape, within a behavior-driven process, organizes and builds knowledge about this consumer language – communicated in not just words, but also imagery and memories. The importance of this knowledge is to provide a basis for building cues in concepts. It is through cues that conceptual appraisal achieves emotional impact. However, cues and their respective emotions must also fit within the guardrails of the product line.

The challenge in understanding this landscape is that cues and emotions arise from the unconscious mind. Most research techniques engage the conscious mind in rational thought. Therefore, the research methods must indirectly measure emotions and their associated cues to truly understand this landscape.

The answer to this challenge comes from the fact that cues are formed through priming – where appraisal during past experiences has sufficient emotional impact to be stored into the memory of consumers. The research methods used in building the scoping landscape involve the recall of past memories and associations to capture the implicit mind of consumers.

There are two approaches to build this knowledge about the link between cues and emotions. The first approach involves an application of online technology that can filter through the vast quantity of online peer-to-peer chatting, to select dialogue about specific brands and product experiences. Implicit in these communications is information about the qualities of foods and associated sentiment (feelings). In this way, the identified qualities are indications of cues that elicit the associated sentiments.

Social dialogue research is advantageous in that it can be a quick and low cost alternative to generate information about cues. However, it is weak in that the sampled dialogues may not come from a target market. Further, it may not pick up on more subtle cues that impact emotions.

An alternative to social dialogue research is to get face-to-face with target consumers using a technique called free association profiling. This technique – while a bit more costly and slow to generate information – not only ensures the right target consumer is being sampled, but it can also pick up on more cues, and more subtle cues too.

Free Association Profiling

Free association profiling was published by Stucky et al.[7,8] as a means to understand the landscape of sensory cues. Free association profiling had earlier been applied only in laboratory experiments, as a method to develop technical language describing sensory qualities in foods. While this particular application was initially developed for fragrances, it has a wide range of applications in emotions research on foods and beverages.

In particular, free association profiling techniques help to understand the language that consumers use to describe product concepts, and to associate that language to cues. It also provides the imagery, memories, or emotions associated with elements of concepts. In these cases, typically more than one concept would be profiled or compared to the rest, or to actual competitive products. The product assessment would be done in the language of the consumer, using their words to describe the different qualities.

The alternative to free association profiling is the focus group or other strictly qualitative techniques. Free association is a hybrid of both quantitative and qualitative techniques. It involves research participants using their own

7. G. Stucky, K. Wiacek, R. DiCasoli, R., Capturing the Implicit Mind: Qualitative Fragrance Imagery by Free Association, in Proceedings from ESOMAR Conference on Sensory Evaluation of Fragrances, May 2005, p. 134

8. G. Stucky, Understanding Utilitarian, Sensory, Social, and Emotional Signals Through Free Association Mapping™. Poster at Pangborn 2009 Conference, Florence, Italy, 26–30 July, 2009

TABLE 8.1 A Comparison of Free Association Profiling and Focus Groups

Free Association Mapping	Focus Group
Individuals develop the language (extensive depth of language created)	**Groups collaborate on language (key descriptors identified)**
Individuals first impressions of products/concepts from every person	**Group influenced impressions of each product**
Uncover relationships that people cannot articulate or rationalize (statistical models)	**Identify only those things that can be discussed or brainstormed**
Trends as understood by research scientist and verified with individual interviews	**Trends as described by the group itself**
Replicable - less groups required for validation	**Highly influenced by moderator and group dynamic**
Language and cultures independent - can compare any individuals or cultures if products are kept constant	**Culture dependent**

consumer terms to describe different aspects of their response to multiple stimuli, and then to rate the relative amount or strength of each quality. It is differentiated from focus groups in how it uncovers associations between aspects of experiences that are implicit in the mind of consumers (Table 8.1).

Consider the following example taken from a case study involving the understanding of the scoping landscape for the instant coffee category.[9] In this case, an actual concept statement for the product line was shown to 12 research participants. Each then developed a number of words to describe the expected product qualities, images, memories, feelings, and anticipated benefits (realized and fulfilled concerns) for this line. This exercise generated 150 different words that were compiled and counted. In addition, it provided a sense for the range and frequency of language used to describe concepts. For example, nine respondents said the line elicited an expectation that would be "creamy" and three said it made them imagine "camping."

---

9. L. Kruse, Assessment of Product Concept Fit Using Free Association Profiling. Presentation at Society of Sensory Professionals, Cincinnati, Ohio, 5–7 November, 2008

However, this analysis did not yet link these terms together – to associate them. To accomplish this task, the research participants were asked to use just their words to describe each of the five instant coffees represented in the line. Further, they were asked to associate words that would characterize the ideal instant coffee characterized by this line concept.

It is important to note that the actual product line does not need to exist – as will be the case in the development of a new product line. What is needed for this technique is to use "protocepts." Protocepts are prototypes or physical product representations of a potential product line. They can come from actual (competitive) products or from products created specifically for the study. These protocepts do not need to actually represent the line concept – only to provide a basis for generating associations.

In this case study, the free association profiling technique resulted in a number of insights into the landscape of consumer language associated with specialty instant coffees. In Table 8.2, an ideal product is described emotionally as a product that makes one feel calm, comforted, satisfied, excited, and invigorated. Whereas, the decaf/sugar-free product was only perceived as calm and relaxing. Associated with a feeling of excited and invigorated were social memories of talking and laughing, and images of places where these memories took place. Relaxing was more associated with non-social memories, such as winter, camping, and reading alone.

This application of free association profiling led to the insight that a concept positioned in the market, as delivering an "excited and energized" experience needs to include cues that elicit social memories through its imagery and words. Insights, such as this, shape the scoping landscape through research.

CONCEPT DEVELOPMENT

The scoping landscape step sets the stage for successful concept development. Having a clear understanding of the language of emotions, and what cues exist to elicit those emotions, is important knowledge that can inspire concept development. Concept development typically involves a process of ideas generation, followed by a process of filtering to generate an idea about each product in the line. There are a number of approaches that an innovation team can take to develop product concepts that fit with a theme. The question of approach requires three decisions: (1) who will be involved in the creative generation of ideas; (2) how will those ideas be filtered down to a set of finalists; and (3) how those finalists will be developed into concepts for the refinement and validation step.

Who Generates the Ideas?

In Chapter 3, the innovation team was defined as the researcher, innovation manager, innovators, partners, and consumer. Most companies in the food

TABLE 8.2 Results of Free Association Profiling for Concept (Ideal Product) against Four Protocepts of the Concept

	Ideal Product	Decaf/Sugar Free (Sweet Beany, Metallic, Butyric, Barnyard)	Mocha (Sweet, Nutty, Cocoa, Cherry)	Hazelnut (Coffee, Burnt, Nutty, Tea/Tobacco)	English Toffee (Caramelized, Buttery, Sweet)
Product	Hot, Creamy, Frothy, Smooth, Sweet, Rich	Creamy, Sweet, Rich	Creamy, Sweet, Rich, Hot, Watery, Artificial, Cheap	Creamy, Sweet, Rich, Hot, Watery, Artificial, Cheap	Hot, Watery, Artificial, Cheap, Dark Chocolate/ Chocolate, Chemical, Bad
Images	Starbucks/ Coffee Shop/ Seattle, Den/ Kitchen/ Living room	Reading Room, Camping	Reading Room, Camping	Reading Room, Camping	Denny's/ McDonald's, Dorm room, Camping, Table looking out window
Memories	Oregon coast vacation, Talking/ Laughing, Home/ Coffee Shop	Winter	Winter	College care package, Dorm Room, Christmas in front of fire	College care package, Dorm Room
Lemonade Emotions	Calm, Comfort, Content, Satisfied, Excited, Invigorated	Calm, Relaxing	Calm, Relaxing, Disappointment	Terrible	Disappointment
Perceived Benefits	Energizes, Low calorie, Makes me feel better	Satiating		Satisfy my sweet tooth, Makes me full, Makes me sweat	Satisfy my sweet tooth

industry do not consider the consumer as part of the team. The consumer's voice is heard through insights generated by the researcher. Some companies consider partners (e.g. marketing research suppliers, design firms, advertisement agencies) as members of the team.

Therefore, the typical format for ideas generation is internal – the innovation team applying their own internal knowledge to generate concepts. This is often an appropriate technique when the team has deep knowledge of the specific concerns, cues, and preferences among a target market and product category. However, this is rarely the case when seeking to develop a breakthrough (disruptive) product.

The second approach is to partner with a company that offers concept development services through creative groups. These companies tend to be design firms comprised of highly creative people able to convert a brief (i.e. concept theme) into a large number of concepts that can be subsequently refined and validated through research. This approach has been applied successfully on a wide range of products. However, this approach is dependent on the creative intuitions of professionals who will have their own ideas about what connects with consumers. These perceptions may not be accurate. The use of creative groups can generate a set of concepts that do not hit the product target.

The third and newest approach is to apply the principles of co-creation to co-design ideas and concepts with target consumers. This approach offers a number of exciting new techniques that overcome the weaknesses of these other two approaches.

Co-Creation

The idea of applying co-creation to design products is a fairly recent one. Manyika et al.[10] highlighted this technique, in *McKinsey Quarterly*, as one of the top eight business trends to watch. Two types of co-creation exist. The first is "distributing co-creation," where the company gives up total control over the product innovation and involves members of their supply chain partners in the process. The most prevalent form of distributed co-creation is that used by food retailers and food manufacturers, widely known in the food industry as co-packers.

The second type of co-creation is the bringing together of innovators, researchers, and consumers to dialogue for the purpose of co-design. This not only brings the consumer directly into the design process, it fits with the emergence of the prosumer seeking brands that will work for them. The chief issue with this technique is in finding target consumers who are also creative enough to consider innovative solutions to building cues into product concepts.

10. J. Manyika, R. Roberts, K.L. Sprague, Eight Business Technology Trends to Watch, The McKinsey Quarterly, December 2007

Innovation teams need to be wary of having a vocal few that do not represent a larger market sample, but are able to unduly influence the creation.

Sanders and Stapers[11] propose screening consumers for one of four levels of creative ability – required to participate as a co-designer with researchers, developers, marketers, and designers. At the first level are people who are motivated by being productive; the second group are motivated by making things on their own; the third want to make something with their own hands; and the fourth group, or the highest level of creativity, are motivated by inspiration.

Prahalad and Ramaswamy[12] provide a good overview on consumer engagement for co-creation in non-food consumer goods industries – e.g. consumer electronics and pharmaceuticals. They define co-creating as the creative collaboration between consumer and the researcher, designer, marketer, or developer for product innovation. Their model is to recruit consumers engaged in seeking behaviors – seeking information about companies, products, prices, consumer actions, and reactions. Their model for co-creation involves engaging these consumers through the building of social networking communities, enabling consumer "experimentation" with developed products. They claim the "building blocks" for this model are dialogue to create loyalty, access to information through brand owner sharing, open risk assessment with focus on reducing harm to consumer, and transparency of information on costs, prices, and margins.

Hottenstein et al.[13] reported the use of "Focused Guidance Groups" as a hybrid that uses elements of both qualitative and quantitative techniques where "consumers explore and quantify consumer liking for a product, service or concept." However, these groups appear to involve only the researcher and consumer.

The other case study comes from Kettle Brand Potato Chips (Chapter 6). Between 2004 and 2008, Kettle Brand (now Diamond Foods) applied co-creation to established ideas for new product. Food writer Jeff Crites[14] posted the following about Kettle Chips.

For the past four years, Kettle has turned over product development to fans, inviting chip lovers to help select new flavors to join the existing line. Inspired by "passionate requests" for the ultimate hot chip, People's Choice IV took a spicy

11. E. Sanders, P.J. Stapers, Co-creation and the New Landscapes of Design, 2008 http://www.maketools.com/articles-papers/CoCreation_Sanders_Stappers_08_preprint.pdf, 2011

12. C.K. Prahalad, V. Ramaswamy, Co-creating Unique Value with Consumers, Strategy & Leadership, 32 (3) (2004) 4–9

13. A. Hottenstein, D. Creighton, S. King, Focused Guidance Groups: a Qualitative/Quantitative Approach for Product Development Guidance. Presentation at Society for Sensory Professionals, October 28, 2010

14. http://www.consumerpassion.com/consumer_passion/2008/01/votes-are-in-ne.html, posted on January 15, 2008. January 10, 2011

turn with a "Fire and Spice" theme and five fiery nominees: Wicked Hot Sauce, Mango Chilli, Jalapeño Salsa Fresca, Orange Ginger Wasabi, and Death Valley Chipotle. Death Valley, with its blend of red chilli, cayenne, chipotle and habanero, finished as the top pick.

In speaking with Chief Ambassador, Jim Green, as cited in Chapter 6, their "People's Choice" program involved an annual asking of their loyal fans to submit ideas for new chip flavors. The innovation team then screened through all these ideas, sorting, and finally developing a set of "most popular" ideas into full concepts. These product concepts were then turned back over to fans to vote on the winner. The program appears to have stopped with the purchase of Kettle Foods by Diamond Foods in 2009.

Filtering Ideas

This leads us to the topic of how to filter ideas so that they can be built into product concepts. There are several factors that contribute to the selection of a technique. The first is a question of the need to filter ideas. Whether there are thousands of ideas submitted by co-creating, loyal consumers, as in the case of Kettle Chips, or 20 ideas generated by the innovation team, there is often a need to filter these ideas down.

A typical consumer research technique applied to filter down ideas is to apply choice-based analyses. Choice-based analyses involve research participants picking winners from sets of ideas or concepts. Various different approaches to choice-based analysis can involve second evaluations by research participants, where the winners from sets are also scored on various quantitative response measures. It is typical to have the "winners" ranked in this second set of evaluations. The result is the type of filtering that can be observed in Figure 8.3.

In this scatter plot, the "Keepers" are the ideas that have been frequently chosen as winners and are ranked or score high in the second evaluation. "Also-rans" have been frequent winners, but score low in the second. "Niche" ideas have not been frequent winners, but tend to score high when selected. The others are filtered out as "Losers." Depending on concept development objectives, it may make sense to keep strong "Niche" ideas or strong "Also-rans".

A second technique that has gained significant popularity is to pick not only the best, but also the worst from each set of products, i.e. typically called a maximum-difference (Max Diff) scaling technique.[15] The winners can further be rated independently or appraised under more holistic contexts. These added pieces of information lead to more criteria within which to filter.

---

15. http://www.sawtoothsoftware.com/products/maxdiff/. January 13, 2011

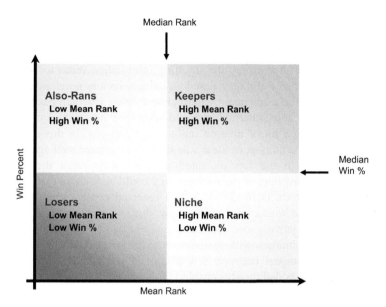

FIGURE 8.3 Quadrants of a choice-based analysis with ranking of winners. (Please refer to color plate section)

An example of a more holistic approach to filtering is to include competitors into the choice-based analysis. A new method[16] has been developed that includes up to three different interconnected choice-based analyses to make the screening process more holistic. This method includes a competitor choice, initial screening of ideas choice, and choices comparing winning competitors and winning new ideas.

In the competitor choice, each consumer picks the products they typically use, or would consider using, if their "typical" product were not available. Winners from the initial screening of ideas are then combined into the third set of choices comparing competitors and winning ideas. By centering and scaling based upon the marginal frequencies for each tournament, the information from all three choice-based analyses can be combined.

Consider the case study of the development of a new dog food concept. In 2010, there were 83 major dog food varieties identified as being sold in US food retailers. In addition, 30 new ideas were developed by an innovation team for a new product concept that could compete within the current marketplace under a new brand. Consumers, as research participants, selected the typical dog food and five alternatives they would consider. They then screened the 30 new ideas into six winners (and three additional runners-up within six groups of five ideas).

16. J. Scheafer, D. Plaehn, Playoff: A Methodology for Testing Product Ideas in a Competitive Context, unpublished. InsightsNow, Inc.

These were finally appraised using a best–worst picking process. The results are shown in Figure 8.4.

Since the basis for choice-based analyses is the selection of a winner (and loser) from a set of alternatives, the resulting information tends to be more emotionally driven than techniques based on ratings, such as purchase interest or intention. When there are many sets to choose from, research participants tend to select from choice alternatives based upon their intuitions (emotions), rather than spending a lot of time on rational comparisons. In this case, the emotional impact of a number of ideas was shown to compare well against the competition, although the top six rated products were from the list of competitors. The sensitivity of the methodology could be seen, in that a major brand with three products in the study had received considerable negative media attention three weeks prior to the study, due to a product recall. These products received relatively low win rates.

Choice-based techniques with competitive context provide insights into the relative emotional impact between new ideas and a consumer's incumbent brand. The resulting added insights enabled the filtering of ideas and concepts, not just on the basis of winning or losing, but on being able to win against specific brands and/or products. The information from this approach to filtering can also be based upon the relative difference between market segments in appraisal of ideas against competition. For example, this could include filtering criteria comparing winning and losing against loyal users of competitive brands; or, in the case where a disruptive concept is being developed, the ability to win against most or all of the incumbent brands in the marketplace.

TURF – Picking the Best Combination

Choice-based techniques involve consumers picking the winning idea out of a set of ideas and forcing a ranking or scoring of winners to generate indexes to filter ideas. TURF (Total Unbiased Reach Frequency) is a standard method that researchers use to assess what is the best combination of ideas for the product line. Choice and TURF can be combined into a single study, since they can be based on the same data.

The typical TURF analysis is based on simple frequencies searching for combinations of ideas that have the highest percentage of at least one winning idea on the combination. In this way, a series of niche products can achieve greater reach than one selected more frequently alone. An exciting advancement in TURF analysis is the application of Shapley values (Chapter 7) to the problem of estimating combinations with greatest reach. Shapley can incorporate multiple response measures. This leads to new opportunities to incorporate metrics generated from emotions research into Shapley values.

Consider the case of a new beverage line, Summertime Taste. Consumers were first exposed to the concept theme, in the form of a product concept "Old-Fashioned Lemonade." They were then asked to submit new ideas for additional

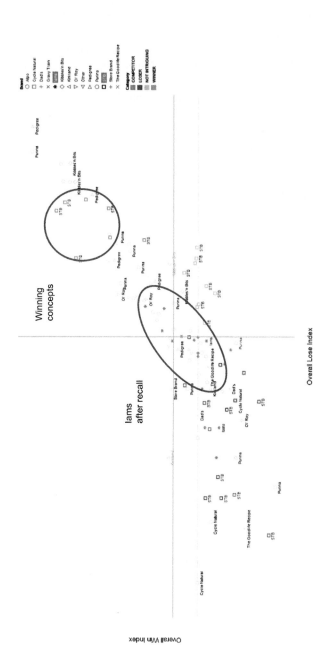

FIGURE 8.4 Results of choice-based analysis using maximum difference scales comparing new dog food concepts (STB) against competitive products. Six test concepts scored well against the top brands. Iams scored unusually low after a product recall. (Please refer to color plate section)

flavors. This resulted in 19 ideas for other flavors that all fit the Summertime Taste theme and might be associated with the "Old-Fashioned Lemonade" product. In this case, a standard choice-based analysis was applied to filter the down ideas based upon the relative combined winning percentage and ranking scores (Table 8.3). The resulting index allowed 19 to be filtered down to 11 based upon getting scores greater than 100 (better than average). However, simply picking the top four concepts (including "Old-Fashioned Lemonade") is not necessarily the combination that achieves the greatest reach across the target consumer group. What is required is an additional analysis that goes beyond concept filtering. A measure of reach is required to scope the best combination.

TABLE 8.3 Results of Choice-Based Max Diff Index Score for Ideas to Extend the Summertime Taste Line of Beverages

Item Name	Index Score
Pink Lemonade	133
Strawberry Lemonade	127
Old-Fashioned Lemonade	126
Mixed Berry Lemonade	112
Pomegranate Lemonade	108
Iced Tea Lemonade	107
Tropical Twist Lemonade	107
Raspberry Lemonade	104
Mint Leaf Lemonade	102
Tahiti Lime Limeade	101
Blueberry Mix In Lemonade	100
Black Cherry Lemonade	99
Strawberry-Kiwi Lemonade	94
Orange Squeeze Lemonade	91
Citrus Supreme Lemonade	90
Acai Berry Lemonade	85
Green Tea Lemonade	75
Grapefruit Squeeze Lemonade	73
Mango Mash Lemonade	72

There are a number of different measures of reach that might be included to augment a typical choice-based analysis. In this case, traditional TURF analysis and three alternative Shapley value measures were applied to provide different insights into the best combination of three additional products to a line that included the must-have "Old-Fashioned Lemonade" (Table 8.4).

Traditional TURF involved top-2 purchase intent scores when the product was rated a winner from the choice-based analysis, saturation, opportunity score, and Shapley value. The strongest combination (overall rank) among all four measures was the combination of Pink Lemonade, Strawberry, and Limeade.

Research techniques covered, so far, include the generation and filtering down of ideas, and to further focus the conceptual design for a whole product line based upon measures of reach. What these techniques do not cover is how to take an idea and build it into a rough, full concept.

This process is actually similar to what has already been covered in Chapter 7. In that chapter, themes were used to define potential solutions for the conceptual definition of the whole product line. In this way, the resulting opportunity solution is the template upon which rough concepts for each product in the line can be developed. The filtered and combined ideas, through choice-based analyses and measures, can be incorporated into this template as the set of concepts for refinement and validation.

CONCEPT REFINEMENT

The process of filtering down ideas and finding the right combination of ideas that characterize a line leads to the formation of a number of candidate product concepts for the whole product line. Each idea (e.g. flavor) is incorporated into the concept template to form a separate concept statement.

Once a number of candidate concepts have been developed, the question then turns to how best to refine and validate these concepts to achieve the goals of the Define phase.

Concept refinement involves the building of cues into concepts to provide meaning and increase emotional impact. Concept cues include those already built into a template from which ideas were generated within the concept development step. Cues can involve specific words, imagery, color, shape, and form of concept statements – that elicit emotional impact during appraisal.

The imagery of a concept is typically communicated through graphic mock-ups of packaging front panels. Relevant emotions within which to optimize concepts depend on how cues have been designed into the concept. In addition, the surface concerns, prior expectations about the respective brand, and context at the moment of an appraisal contribute to how those cues elicit emotional impact.

A number of research techniques are proving valuable in providing emotions insights to optimize cues. Some techniques involve the application of new

TABLE 8.4 Results of TURF, Saturation, Opportunity and Shapley Value Differences for Three Product Line Combinations

Combinations	TURF Reach		Saturation		Opportunity Score		Shapley Reach		Overall Rank
Pink, Strawberry, Limeade	52.7%	(1)	28.9%	(3)	192.3	(3)	46.3%	(6)	1
Strawberry, Limeade, Iced Tea	52.6%	(2)	27.0%	(25)	79.9	(48)	42.1%	(27)	21
Strawberry, Limeade, Raspberry	52.2%	(3)	27.9%	(10)	194.9	(2)	43.0%	(22)	3
Strawberry, Limeade, Tropical Twist	52.1%	(4)	26.7%	(32)	88.5	(42)	41.9%	(33)	23
Pink, Limeade, Iced Tea	51.9%	(5)	27.0%	(25)	165.0	(9)	41.9%	(30)	10
Pink, Strawberry, Pomegranate	51.8%	(6)	27.6%	(16)	169.9	(7)	45.3%	(9)	4
Strawberry, Limeade, Pomegranate	51.8%	(6)	26.3%	(45)	137.1	(23)	40.8%	(46)	26
Pink, Strawberry, Raspberry	51.6%	(8)	27.9%	(10)	160.0	(11)	47.5%	(2)	2
Pink, Limeade, Raspberry	51.6%	(8)	29.3%	(1)	141.7	(21)	42.9%	(24)	7
Pink, Raspberry, Mint Leaf	51.5%	(10)	27.7%	(12)	207.1	(1)	42.9%	(23)	5

technologies to understand the relevant attention given to specific elements in a concept. Some techniques assess the respective meaning and emotional impact of that element.

Highlighter and "Mouse-Over" Techniques

Highlighter techniques are proving to be a great way to gain insight into what words and phrases, within concept statements, are cues and how they elicit emotional impact. They involve consumer appraisal of concept statements, using computers where participants are asked to highlight words and phrases of meaning to them. This is typically followed up by a series of open-ended text boxes to comment on the meaning of each item highlighted. A variant of this is to apply "mouse-over" or "click-through" techniques, to identify what parts of images and pictures are attention-getting, followed by open-ended text boxes to capture the meaning of specific elements of imagery and pictures.

An application of the appraisal framework is to break down responses into a more emotive basis – i.e. inferred information about what were the underlying associations, concerns and emotions to highlight or mouse-over concept elements. This type of text analysis can provide deep insights into the relative emotional impact of different elements designed into a concept statement. These types of insights can be valuable in the refinement and optimization of elements to maximize emotional impact.

The relative value of these techniques is that they can easily be incorporated into Internet surveys to target consumers as research participants. As a result, these techniques lead to rapid research iterations involving large numbers of consumers at relatively low cost.

Eye-Tracking

Eye-tracking is a powerful technique that has a wide range of research applications.[17] It can be applied to concept research to understand what elements of a concept elicit attention. This technique is based upon technology that tracks eye movement and gaze from research participants viewing content displayed on specially built computer screens. These typically involve video cameras that pick up reflections off the cornea of the eye from infrared light emitting diodes at specific frequencies.[18] The tradeoff of eye-tracking, in concept refinement, is in its requirement for special hardware that limits its application to central location testing. This slows the

---

17. http://en.wikipedia.org/wiki/Eye_tracking

18. P. Olsson, Real-time and Offline Filters for Eye Tracking. MS thesis, KTH Electrical Engineering, Stockholm, Sweden, 2007

speed in data collection and increases the relevant cost of concept refinement research.

Eye-tracking has the advantage that it measures not only what elements of a concept statement are grabbing a consumer's attention, but it also tracks physical metrics such as eye pupil dilation. There is some evidence that eye pupil dilation is associated with direction of the emotional response. However, assessing eye dilation in response to visual stimuli has proven difficult, due to confounding from visual content that varies in color and brightness.[19]

Neural Measurement

Neural measurement techniques are beginning to find their way into the portfolio of research techniques for product innovation. Neural measurement techniques involve the capturing of neural activity as a more direct way to measure emotions than others covered, so far, in this book. These techniques are still highly experimental, but offer promising opportunities to measure emotional impact that is more deeply imbedded into the unconscious mind of consumers.

Two startup companies – Emsense[20] and Neurofocus[21] – are currently offering solutions. Both offer research services based upon proprietary wireless, non-intrusive headgear that research participants wear, with data collected as the participants appraise concepts they are exposed to on the computer screen.

Both solutions offer the advantage of being mobile, enabling research participants to appraise concepts elements within a more natural context, such as their home. However, these solutions require the cleaning of artifact data captured in the course of research that is unrelated to the conceptual stimuli. In this way, they are slower and more expensive than other concept refinement techniques. They claim a greater accuracy in gauging emotional impact than other techniques. However, while promising to get closer to measuring the implicit mind of the consumer, they do not deliver insights on their own for concept refinement. They require the integration of other information generated at the point of appraisal.

19. L. Granka, K. Rodden, Incorporating Eye Tracking into User Studies at Google, http://www.google.com/research/pubs/archive/34377.pdf, Proceedings of ACM CHI 2006 Workshop on Getting a Measure of Satisfaction from Eyetracking in Practice. January 15, 2011

20. EmSense Announces Release of Revolutionary EmBand24™, http://www.emsense.com/news/electroencephalography-EEG-brainwave-measurement.php. February 15, 2011

21. http://www.neurofocus.com/pdfs/NeuroFocusExecutiveBrief_BeverageTCE.pdf. February 20, 2011

Choice-Based Conjoint Techniques

One way to integrate additional information into the concept refinement process is to apply choice-based conjoint.[22] This involves using the template generated in the building of concepts and altering the elements of concepts according to an experimental design. As was described in the section on opportunity refinement (Chapter 7), this same technique can be applied to identify the optimal combination of elements.

Choice-based conjoint techniques involve consumers appraising a series of different concept statements – and then selecting winners from subsets according to factorial experimental design. This experimental design results in the estimation factors from each research participant. The estimation factors are the main effects and limited set of interactions used to identify the optimal combination of concept elements that will maximize choice. These estimates are combined to estimate the optimal combination of the target consumer or different segments.

Since these techniques are choice-based, they involve less cognitive assessment than results from ratings-based conjoint techniques. This implies that they offer greater prospects to be more emotive-based. The tradeoff of choice-based conjoint techniques is that they do not guide concept refinement through insights into the "whys" of choice. No information other than the change of elements is provided. This requires the addition of information beyond the design elements, and information on how consumers react to concepts presented as a whole.

Protocepts

As was discussed Chapter 5, the value of the appraisal framework is in how it can be applied to integrate different information sources into insights. Each of the above research techniques has its relative tradeoffs as an emotions measurement technique for concept refinement. Some techniques are more cost effective, some lead to faster delivery of insights, and some are more easily integrated into insights. It is this third category that leads to the discussion of protocept techniques.

Protocepts involve the appraisal of a common prototype product as the concept varies. These involve situations where a "benchmark" product (proto-cept) can be used to provide an experience outcome that is dependent on the expectations generated through the concept appraisal. The incorporation of protocepts provides a basis to optimize which cues are built into the concept, on the basis of how they form expectations to translate to the product experience.

---

22. K. Chrzan, B. Orme, An Overview and Comparison of Design Strategies for Choice-based Conjoint Analysis, http://citeseerx.ist.psu.edu/viewdoc/download?doi=10.1.1.87.597&rep=rep1&type=pdf, 2000. January 29, 2011

Integrated research designs, as have been discussed in earlier chapters, involve a more holistic approach to appraisal where research participants iterate through different aspects of the product experience. These same types of designs can be applied to concept refinement when test concepts are assessed, followed by the consumption of a fixed food protocept.

Consider the case study for a microwavable cake mix. A food company was looking to develop the concept for a simple way to prepare dry mix that would result in the highest expectations of an authentic food when microwaved. As no real microwavable cake mix existed, a holistic, integrated research design needed to be implemented. In this case, the concept statement varied in what other ingredients had to be combined with the dry mix prior to microwave cooking. Twelve different concepts (including simple instruction statements) were each presented to research participants in a monadic study. Participants read the concept and prepared the product using ordinary kitchen ingredients. The prepared mix was then taken away for microwave cooking, and returned for tasting and assessment.

The assessments occurred at concept, preparation, and consumption. The goal was to see how the concept resulted in increasing overall desire to repeat using the product, if it were available. What was found was a 35% swing in the desire to purchase the product after preparation. Added complexity in instructions, by including more ingredients, resulted in less interest in seeking the product in the marketplace, if it were available. A few ingredients – when required to be added to the mix – were found to signal a feeling of greater quality and authenticity as a more homemade food. There was also 15% swing in the purchase interest after consumption, even though all products essentially were identical. The influence of concept and instructions impacted the overall product desire (see Figure 8.5).

CONCEPT VALIDATION

Concept validation techniques involve the testing of fully developed concepts against success criteria associated with concept design goals. These criteria are often key decision criteria in building a business case, backed by corporate resources, that supports the innovation team's activities in the next phase of product development – product design. The decision metrics used in building a business case typically include metrics about market size or volumetric forecasts. Concept validation goes beyond volumetric, market response metrics – it includes behavioral metrics and benchmarks.

Price Sensitivity to Concept Elements

A key decision metric, important in the building of a business case, is the perceived price–value tradeoff for a product. Often the category within which a product will be introduced will lead to price point as a point-of-entry

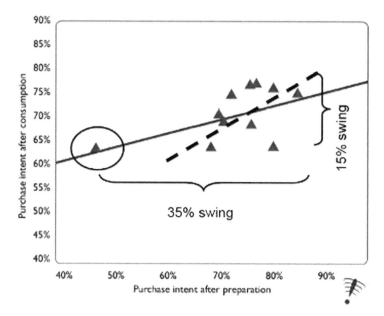

FIGURE 8.5 Relationship between preparation and consumption scores on protocepts for dry mix food product.

requirement. Such is typically the case for incremental innovations, such as a line extension or flanker product line. However, the price point may not be so readily known for more innovative products seeking to disrupt the category.

There are two approaches to understanding the price point for a product concept. The first is to incorporate price into a choice-based conjoint study. The alternative is to use pricing as a response variable, rather than a design element, as in conjoint. The most traditional technique is the Van Westendorp approach to price sensitivity.[23] This technique asks research participants to rate what pricing point would be too expensive to buy, too cheap to consider of sufficient quality, expensive yet they would still consider, and too inexpensive to consider a bargain. These responses provide the basis for a probabilistic estimation for a product's price point (see Figure 8.6).

The application of the Van Westendorp approach to concept validation involves the use of this technique as the front end to an integrated research design. In this application, the response points – where a product is too cheap or too expensive to purchase – are used to generate choice alternatives between a benchmark or incumbent brand and a test concept. This approach requires the real-time calculation of the price points for each research participant. These are

---

23. P. Van Westendorp, NSS-Price Sensitivity Meter (PSM) – A New Approach to Study Consumer Perception of Price. Proceedings of the ESOMAR Congress, 1976

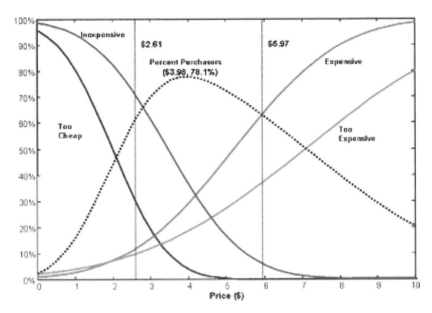

FIGURE 8.6   Van Westendorp approach to concept validation by using pricing as a measure of price-value. (Please refer to color plate section)

then used to generate actual purchase choice sets from which participants must choose.

Validating Disruptive Potential

As discussed earlier in this chapter, the success of a product requires both disrupting the use of an incumbent and introducing an alternative. This leads to the question about how to gauge the ability of a concept to be disruptive. Can a disruption metric be developed to validate concepts?

Consider again the pet food case study where unusually low win rates for a leading dog food brand were believed to be the result of a product recall. Disruptive products can have similar impact on their competitors. They introduce new concerns about the delivered competitive product experience. The higher the ranking in emotional impact from such a concern (e.g. heightened concerns that can be associated with more basic need states) the lower the cost of entry for a new product. Consider each of these four disruption strategies:

(A) Heighten existing functional concerns into the healthiness (survival fears) of certain dog food products with the introduction of a new product that is the solution to this concern.
(B) Heighten social fear of rejection among dog owners belonging to social networks by associating membership with use of special "ultra-premium" dog food products – such as a new product.

(C) Increase hope for the self-actualization for your dog by offering a new dog food that cues an authentic premium food experience for dogs.

(D) Increase hope that a dog's happiness will be increased by offering a line of intriguing varieties that are similar in taste to different types of "human table scraps."

In the case of these four alternative disruptive strategies, strategies A and B are examples of concerns that have the highest potential for disruption on an emotional impact basis. However, functional differentiation is typically more difficult to protect and is often associated with incremental – feature enhancing – innovation strategies. Therefore, strategy B offers perhaps a greater chance for differentiation on the basis of social concerns.

On the other hand, strategies C and D lead to concerns based upon self-actualization need states. The anthropomorphic transfer of human to dog happiness based on intriguing human table scraps may be novel, but offers a much weaker basis for disruption. These sorts of strategies may be better suited to the introduction of a new product that is carving out its own niche – not requiring the displacement of an incumbent.

The measurement of this type of disruption requires a head-to-head comparison. One approach might be to conduct a survey comparing the test to an incumbent. Another might be to have each concept (incumbent and test) compared in monadic fashion. However, neither of these techniques provides a basis to compare the full level of disruption of new concept against the incumbent.

This is where the idea of a protocept offers a more holistic approach to research. A monadic cell, with concept tested immediately after the use of the incumbent, provides better context within which to test the emotional impact of a new concept. Likewise, a monadic cell, with the concept tested immediately before use of the incumbent, provides better context within which to test the impact of the test concept on the incumbent. Comparing the relevant differences in appraisal measures of emotional impact for the test concept and incumbent provide a great test to validate disruptive potential.

These concept design and validation techniques serve to provide the innovation team with product concepts that have a strong chance for success. This is the "good" news. The "bad" news is that the product must now deliver on the promise of the concept. This leads to the second half of design – product design.

Key Points

- The concept is the heart and soul of a product. It is the target that all innovators on the innovation team use to design and build harmonious products.
- The conceptual appraisal framework extends Maslow's need states into a hierarchy of surface concerns and emotions for disruptive product innovation.
- The goal of Define for breakthrough innovation is to change behavior, by both disrupting existing habits for incumbent use and generating trial for the new product.

- The guardrails for concept design depend on the behavior drivers, cues, and emotions implicit in the theme discovered to be the opportunity.
- The process for Define involves developing the concept landscape, and using this landscape as the knowledgebase upon which to develop and refine concepts.
- Concept landscapes are generated from consumer product insights into the expected qualities, images, memories, feelings and anticipated benefits associated with the theme for a new product line. Free association provides an alternative to focus groups to understand these associations implicit in the unconscious mind of consumers.
- Early iterations of concept development are best facilitated by dialogue between consumers and innovators, through co-creation methods, to generate ideas to fill out the variable part of concept templates.
- The filtering of ideas is facilitated by incorporating competitive products into choice-based tournaments — helping identify ideas that can both disrupt the incumbent and lead to trial for a new product idea.
- The best combination of ideas can be selected to achieve the broadest reach in emotional impact over the target consumer through Shapley values and other metrics.
- A number of emotion-based measures can be applied to refine concept, including choice-based conjoint, highlighter and/or "mouse-over" tools, eye tracking and neural measurement.
- Protocepts offer a more holistic approach to refine concepts by measuring the gap between the promise of the concept and a benchmark prototype experience.
- Concept validation can be achieved by incorporating price as a gauge of price-value sensitivity or by testing the disruptive impact of a concept on a habitual behavior.

Chapter 9

Design

Colors, like features, follow the changes of the emotions.

Pablo Picasso

TRANSLATION

The concept may be the heart and soul of products, but it is food product design that delivers emotional impact. Food product design involves translation of the expectations generated by a concept into a product experience. While conceptual design involves building cues and meaning into concepts to elicit anticipation emotions, food product design involves building cues and meaning into the package and packaged food to elicit both anticipation emotions and experience emotions.

Food product designers are the package designer, package engineer, research chef, and product developer. They contribute to innovation teams as domain experts that apply their knowledge to create the experience. A recurring theme throughout this book is that a key determinant of the experience is the context within which a product appraisal occurs. For this reason, considerable attention is given in this chapter to interaction design – a more holistic approach to design that takes into consideration how the consumer, package, packaged food, environment, and situation interact.

Contemporary design is quickly moving to a highly collaborative model between consumer, researcher and innovator (i.e. designers). This chapter provides an overview of research methods that are highly collaborative and that are used to inspire and guide food product designers. It focuses on iterative research methods that deepen insights into how to achieve emotional impact through the iterative creation and testing of mock-ups and protocepts. Mock-ups are early design phase representations of packaging. The use of the term "protocept" is defined differently here than in Chapter 8, where we looked at conceptual design. In concept design, protocepts were defined as "physical product representations of a potential product line." In food product design, the concept is fixed with "prototypes" varying. These protocepts are used to guide the translation of the design into an experience.

Breakthrough Food Product Innovation Through Emotions Research. DOI: 10.1016/B978-0-12-387712-3.00009-8

The outcome of the product design phase is a set of product requirements for product development to build a commercial product. These requirements include the sensory qualities built into the package, and packaged food that generate cues signaling what experience to expect. They also include the qualities of products that deliver the experience. This translation of concept into experience gets the innovation team closer to the target – the destination to which the brand will take trusting consumers.

PRODUCT DESIGN GOALS

The goal of food product design is to change consumer behavior, by impacting consumers on an emotional level during product experiences. This goal is achieved by translating the concept into a set of product requirements. These requirements include the design elements of the package and the intrinsic qualities of the packaged food itself.

The goals for package design are to impact the product experience indirectly and directly. Packaging communicates the product concept, forming anticipation emotions when a package is appraised within a shelf set or a pantry. These emotions motivate choice (i.e. selecting behavior) and create expectations about the consumption experience.

The design elements for the packaging and intrinsic qualities of the packaged food both contribute directly to experience emotions through use and consumption. These anticipation and experience emotions provide the basis for behavior change among target consumers. Conceptual design impacts the behaviors of seeking (e.g. disrupting habits leading to a search for an alternative product) and selecting (i.e. new product trial). Food product design ultimately results in a delivered experience – impacting the behaviors of sharing (i.e. telling peers about product experiences) and sensing (i.e. repeating trial through the formation of new habits). The package, however, plays a dual role of contributing to selecting behavior.

Therefore, the innovation strategy – the behavioral strategy for how the brand is to win in the markets where it will play – is an essential determinant of the goals for the product design phase. While it is the concept that breaks habits and leads to new product trial, it is the experience itself – generated through food product design – that leads to a desire for a consumer to repeat the experience, forming new habits.

Guardrails

Food product design is a highly creative process. Therefore, it needs guardrails in order to remain focused on the experience target. Guardrails start with an understanding of the design landscape within which the consumer lives – the cues in the environment that motivate behavior, cues that have been built into competitive products, and cues that might be built into a new product.

As with discovery and concept design, specific guardrails are set up depending on innovation platforms, e.g. a technology upon which the product design must be built. Likewise, if the innovation is market-driven, then the guardrails may serve to restrict design elements to specific regions of the behavioral landscape. As with earlier phases, the landscape associated with the opportunity solution provides the knowledge base for the product design process, including cues that may or may not fit with the concept.

This knowledge helps the package designer deliver functionality and/or aesthetics through cues designed into shape, form, closures/openers, hand holders, and a wide range of design elements. The communication or messaging on the package panels provides cues that deliver meaning, including self-identity, social identity, and/or psychological intrigue. The packaged food itself has cues built into it that signal much more than sensory emotions, such as enjoyment and liking. Colors and aromas can signal health-related fears, or cue a food as appetizing and/or authentic. The mix of cues (i.e. instant specialty coffee dry mix and foaming action) can elicit amusement.

Within this landscape, the guardrails for a food product design are further established by the product concept itself – i.e. how the product will be known, the branding, positioning, product claims, and features. Because the concept forms the guardrails for what expectations the consumer has about the experience, it constrains how the packaging and the consumable contents can be designed.

How cues are built into the design requires continuous consumer feedback. As a result, food product design is highly iterative, especially in the early design stages. Early design often involves building rough protocepts and mock-ups in laboratories. Therefore, much of the consumer feedback to experiences occurs within controlled laboratory environments. In this way, early stage design must rely on simulated experiences, based upon what is known about the design landscape. Only by getting out of the laboratory and into real context can the real impact of food product design be known.

APPROACHES TO DESIGN

The importance of design became a key rallying point for the CPG industry-leader P&G, under its CEO A.G. Lafley. A *Business Week* article, in 2005,[1] described how Lafley formed a board of industry design leaders to provide input on P&G's product development and marketing strategies. In addition, Lafley elevated design through the creation of a new executive position. Claudia Kotchka (P&G VP of Design, Innovation, and Strategy) stated, "Functionality is not enough. We want to identify consumer desires, rather than needs. What gives you the 'wow'."

1. Robert Berner, P&G Quest for 'Wow' Design, Business Week, August 1, 2005, Special report – Get Creative/Online Extra. http://www.businessweek.com/magazine/content/05_31/b3945423.htm. January 5, 2011

P&G's reliance on design to achieve emotional impact (i.e. "wow" impact) highlights its focus to disrupt. By 2005, P&G had quadrupled its design staff, implementing a corporate design focus on generating emotions. Designers were incorporated into product development, working with the R&D staff at the beginning of the build process to help conceive products.

This approach to product development, through design, has implications for the food industry. It highlights the importance of adding design into the innovation and product development process for food companies. It serves to take the innovation focus away from functionality, through features and sensory appeal that increase hedonics. It places more focus on sensory holistically – not just an avenue for making a food taste good, but more importantly a key aspect in every phase of the food experience through cues.

Designers approach package and packaged food development differently than the food science trained product developer. While the food scientist (product developer, food engineer, or packaging engineer) is trained in the scientific method, the package designer or research chef is trained in the art of design. As discussed earlier, the scientific approach uses an atomistic (reductionist) approach to learning. The designer and chef both embrace the holistic approach to learning.

Interaction Design

The holistic approach to design is an essential step in the product development process. It serves to overcome an important challenge in food product design – translating the promise of the concept into an experience that delivers achieve emotional impact. This challenge is complicated by many complex interactions between the consumer, product and context of use. The creative designer requires insight into these complex interactions to overcome this challenge.

Insight into this complexity starts with understanding that the landscape of behaviors, behavior drivers, and cues is complex within a consumer target. There are different types of interactions of importance in packaging and packaged food design. While most innovation teams focus on the interaction between the consumer and the food product, there are also important interactions between the consumer and the situational context, within which product experiences occur, e.g. change in mood state. There is also the interaction between environment and the food product. This includes how cues might be masked or altered by the environment, e.g. color masking due to light in the environment.

Interaction design also includes behavior. Jodi Forlizzi,[2] at Carnegie Mellon University, presents a model of how interactions among products, consumers, and contexts lead to change in sensing and sharing behaviors (see Figure 9.1).

---

2. J. Forlizzi, The Product Ecology: Understanding Social Product Use and Supporting Design Culture, International Journal of Design 2 (1) (2008) 11–20

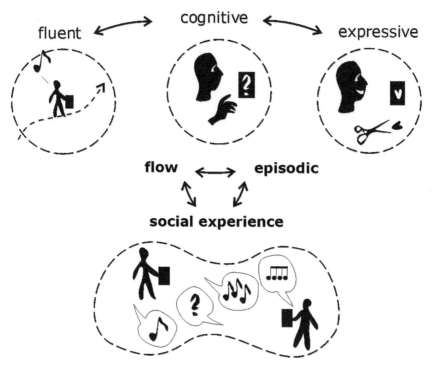

FIGURE 9.1 Design model of how consumers change over time in their interactions with products, and how those interactions in turn lead to the behaviors of sensing and sharing. *(Source: J. Forlizzi, "The Product Ecology: Understanding Social Product Use and Supporting Design Culture," International Journal of Design, 2008, 2 (1), pp. 11–20. http://www.ijdesign.org/ojs/ index.php/IJDesign/article/view/212)*

She describes three behaviors that consumers normally cycle through. The first, the "fluent" (flow) consumer behavior, is identical to sensing (habitual) behaviors. When this experience is disrupted (episodic experiences), the cognitive mind takes over. This tends to lead to expressive (sharing) behaviors – social experience.

These experiences can be viewed in terms of the consumer interaction with the food product. The "fluent" user–product interactions are not competing for our attention. Cognitive user–product interactions focus consumer attention on the product at hand. They allow consumers to focus on the consequences of food preparation of consumption activities. Expressive user–product interactions help consumers form relationships to a food product, or some aspect of it. In expressive interactions, consumers may change, or personalize the food product – i.e. investing effort in creating a better fit between their selves and the product. The stories expressed by consumers to their peers, in turn, lead to forming and changing peer-to-peer relationships.

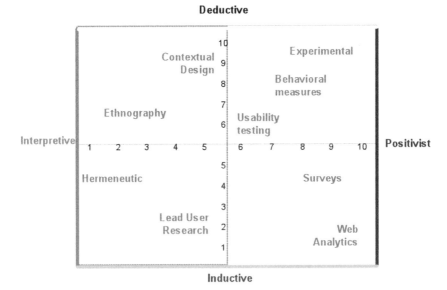

FIGURE 9.2   Categorization of approach to design based upon research methods – inductive or deductive, positivist or interpretive.

These types of consumer–food product interactions lead to new ways of thinking about packaging design and packaged foods. Disruptive foods require that the interaction between consumer and product changes from fluent to cognitive to achieve trial, and then from cognitive to expressive to increase market dispersion. As a new food becomes the new incumbent in the life of the consumer, the cycle reverts back to fluent.

Design Methods

The complexities of the many interactions impacting the food product experience challenge the food product designer. Peter Jones (2009[3]) provides a good overview of various design methodologies that are used throughout the industry. Designers are guided by a number of different research methods (Figure 9.2). These methods range from deductive-positivist techniques, such as might be used with a more scientific approach (e.g. applying experimental psychology or behavioral psychology), to more inductive–interpretive methods, such as the use of "lead users" or observation-based ethnography. He also lists the use of more inductive–positivist methods, such as social-dialogues (i.e. Web analytics).

---

3. Peter Jones, Transforming Contexts in Design Research, http://www.melodiesinmarketing. com/2009/07/25/transforming-contexts-design-research-thinking-peter-jones-presentation/, 2009. February 9, 2011

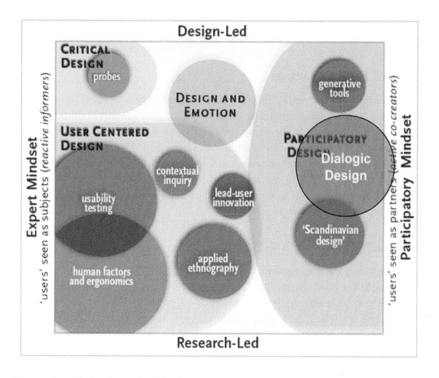

Topography of Design Research. Liz Sanders, Design Research in 2006. *Design Research Quarterly.*

FIGURE 9.3 Categorization of approaches to design based upon who participates (consumers or experts) and who leads the design effort (designers or researchers). Within the more participatory design methods there is an emerging area termed "dialogic design". (Please refer to color plate section) *(Source: Peter Jones, "Transforming Contexts in Design Research," http://www. melodiesinmarketing.com/2009/07/25/transforming-contexts-design-research-thinking-peter-jones-presentation/)*

Liz Sanders[4] provided another way to characterize approaches to design (see Figure 9.3). She makes a distinction between methods that are design-led (e.g. design and emotion) and research-led (e.g. applied ethnography). She also distinguishes between who participated in the design process (the consumer or expert).

Food product design is rapidly embracing a more dialogic design process involving dialogue among the consumer, designer, and researcher. Dialogic design methods can be further characterized on the basis of being more generative and iterative (e.g. co-design with groups of consumers), or more structured and strategic, such as using a small group of consumers to design

---

4. E.B.-N. Sanders, Design Research in 2006, Design Research Quarterly 1 (1) (2006) 3–8

through scenario building ("what-if" discussions) and sense making (i.e. the iterative creation of food protocepts or artwork as mock-ups).

Stages of Product Design

The product design phase is similar to other phases of product development, in that each uses a base of knowledge to inspire divergent (creative) thinking and then to guide convergent (decision making) thinking. In the case of product design, research can sometimes be conducted to extend what is known from the opportunity landscape and to build a more "granular" product design landscape. This is followed by iterations for the creative development of ideas into design elements. Finally, these design elements are validated by testing mock-ups and protocepts with consumers.

PRODUCT DESIGN LANDSCAPES

As with concept design, a more focused landscape is required to target creative development for product design. This landscape is bounded by guardrails and serves the purpose of being the relevant knowledge base within which to design the package and packaged food (or beverage). In the case of the package design, the front panel is a chief determinant of what anticipation emotions motivate selection. The brand, imagery, and key words on the front panel serve to both communicate and cue consumers – motivating trial. The impact of the front panel, along with other marketing communication, also serves to create expectations. These expectations might be in the form of anticipating emotions or perhaps a change in mood, or the anticipated fulfillment of important concerns.

The product design landscape needs to cover these areas, in order that those involved in package design can creatively consider how to build into the front panel cues that both motivate trial and raise the bar high enough, such that the experience leads to emotional impact across the target market. The package itself and the packaged food must also be designed to deliver against expectations, taking care to not raise the bar too high such that expectations are dashed and emotions, such as disappointment, set in.

Landscape Granularity

It is important to understand the level of granularity, or detail, required about consumer experiences in order to begin product design. Elizabeth Sloan, a writer for *Food Technology*, talks about trends in food design in her article "The Pleasure,"[5] where she characterizes a number of destinations that consumers are seeking. She notes that right now people are interested in

5. E. Sloan, The Pleasure, Food Technology 7 (9) (2009) 18–27

"pleasure" in their cuisine. Her definition of pleasure, using the lens of the appraisal framework, is actually a call to build into products more emotional impact from the psychological dimension of the product experience (refer to the Emotions Insight Wheel in Chapter 4). She claims that consumers are "looking for additional excitement in the foods they eat at home. Americans are cooking for fun." They want to "experiment with more cuisines and flavors." There is a "growing sense of boredom," especially "when it comes to fast food restaurants." There are other segments that seek foods that are comforting. She concludes by recommending that design solutions fit into the following four categories:

1. "Treat thyself foods," including treat foods and ways of preparing foods for treating yourself
2. "Comfort Zone" foods that take consumers back to times past or traditional foods that connect them to family
3. "Party Foods"
4. Food ideas for "Cooking Enthusiasts," the people who like to create new foods at home

These design solutions are great examples of experience destinations that characterize different areas of the opportunity landscape for companies. The landscape for food product design is more granular than an opportunity landscape. It focuses on one of these opportunities, and seeks to characterize the cues and symbolic meaning behind the attributes of packaging and packaged foods that might be built into product design.

Consider the organic cereal marketed by Archer Farms. Its cereal product line is positioned as an all-natural, organic food from a brand that supports sustainability. To be differentiated in the cereal category, their product must stand out not just through its name and brand meaning, but also in what else the package communicates through its design.

A trip to the grocery store reveals a lot about the design landscape. It provides immediate information as to how other competing natural, organic cereals are seeking to be different in the minds of consumers. It is these sensory cues that truly differentiate products and deliver emotional impact against concerns and expectations at an appraisal moment. This includes, for example, the fact that the Archer Farms cereal line was distinct, in 2011, by including a front panel window into the package design. This window made visible the color, shape, and texture of the foods inside. While the brand cues the product as for people who support a specific set of lifestyle and life values, the window cues, among other things, "transparency" – revealing the brand is both innovative and has nothing to hide.

If Archer Farms were to seek out new ways to differentiate its line of cereals, it might consider (for example) developing a line of cereals that are perceived as Comfort Zone foods. If this were the case, then the opportunity, concept, and product design landscapes would each respectively need to focus on more defining levels of granularity within the Comfort Zone landscape. For example, a focus on extending this line into the Comfort Zone category would require a different product design landscape than if it did not. The landscape would need to include sensory cues associated with other comfort food experiences. This may require research into different categories of Comfort Zone foods (e.g. chilli, soups, chocolate). The sensory qualities of these foods might also be evaluated (recipes, ingredients, various hot or cold beverage flavors). The packaging of other Comfort Zone foods might provide added knowledge into how a package might communicate comfort and elicit anticipation emotions associated with the expectation that the product be comforting.

In this way, the relevance of "comfort" within the design landscape is as a behavior driver (an expectation), rather than as a feeling (an emotion or mood) to be elicited. Comfort Zone foods, as defined by Sloan, must take consumers back to times past or be traditional foods that connect them to family. The library of cues, within the design landscape, must help package designers create the front panel that triggers imagery of past times or traditional family occasions. However, the design landscape cannot stop there. It must also have, in its library, cues that inspire ideas for how to formulate the packaged food – also triggering imagery of past times or traditional family occasions.

This leads to the discussion about knowledge. How can the researcher bring in the voice of the consumer to provide sufficient granularity inspiring the development of package and packaged food design? The answer to this question starts by understanding the language of consumption and use. Before designers can begin their concept translation, mapping the consumer language of the concept (the promised experience) serves to delineate the attributes of the product through the voice of the consumer.

The Language of Consumption or Use

Food product design is oriented to the consuming or using experience. The language of the consuming or using consumer does not typically involve "purchase interest." In fact, if the product concept is successful in motivating seeking and/or selecting behavior (i.e. trial), then the language used to characterize the product should go beyond "liking." Liking is an expression of product love – a projected feeling onto the product as an object. A product can be liked for its functionality, how it helps build a consumer's social or self-identity, how it stimulates or surprises, or how it leads to sensory pleasure. Since a product can be liked for many things, the use of Liking as a measure is highly flexible. It can fill in when no other measure exists. However, Liking is fairly non-descript. It does not help the researchers or designer learn to develop the design landscape.

In trying to translate the promise of the concept into a product experience, one must go beyond liking and purchase interest to understand sensory cues. This starts with understanding the imagery, memories, and feelings associated with concepts, and how this language translates into sensory perceptions. Once discovered, it is this language that leads to new consumer response measures guiding the development of product design.

Free Association Profiling

In Chapter 8, free association profiling was described as a technique for understanding the language of consumers as associated with specific behavior drivers that characterize the product experience. This same technique can be applied to the product design phase. In this case, the technique is used to develop the product design landscape, with the intended result to inspire creative thinking about possible product requirements (cues) that might signal the delivery of the promised experience (i.e. the concept).

Consider the following case study involving chocolate.[6] In this case study, four alternative chocolate formulations were used to understand "contentment." A group of 12 research participants were recruited representing the consumer target audience. They each evaluated four different types of chocolate through the following procedures:

Step 1 – Warm-up. The moderator explained the flow of the session, then used a warm-up exercise to get the participants thinking about all aspects of a product and its purpose from a utility, sensory, social, and emotional expectation perspective. A paper grid was used with categories that gave direction for associating words and phrases with the product. During this warm-up exercise, group discussions were used to ensure all participants understood and were able to complete the exercise. This took about 10 minutes to complete.

Step 2 – Generate vocabulary. The participants then worked on their own, at their own station. They were shown products either side by side, or sequentially. As they viewed, smelled, tasted, and touched each product, they generated words and phrases that they associated with the experience and with their expectations of use cases. As they associated a word or phrase with the product(s), they categorized the words where they best fit. They were allowed to customize the categories used to ensure meaningful interpretation and effective communication of the product experience.

Step 3 – Rate products. Participants re-grouped and were taught how to use scales to rate their experiences, using their words and phrases. The scale used was "Please rate how much you associate each of your words with this product." Participants went back to their individual stations and were shown products (typically sequential monadic), then scored the degree of association of each of their own attributes for all products, adding new attributes, if needed.

---

6. G. Stucky, Understanding Utilitarian, Sensory, Social, and Emotional Signals through Free Association Mapping™. Poster at Pangborn 2009 Conference, Florence, Italy. 26–30 July, 2009

Step 4 – Interview. By monitoring participant vocabulary and the degree to which they used those words to differentiate the products, individuals were selected in real-time to participate in one-on-one interviews. Interviews focused on understanding the words and phrases the person used, as well as, provided an avenue for the participants to discuss the differences they experienced during the session.

The language and respective language used by participants was analyzed and integrated into insights. Procrustes analysis[7] was used to create a consensus configuration that best associated the attributes with the greatest differences between the products. The results were used to construct multidimensional maps that overlay the attributes of the products. The loadings of the products were mapped against the correlations of each attribute, for each respondent. Consensus and individual maps were used to understand the functional, social, and emotional differences between the products. Analysis was conducted separately for each category.

In the chocolate sampling example, this method resulted in two distinct consensus maps. From the product characteristics (Figure 9.4), it is clear that the major sensory qualities differentiating these products was in their packaging (shiny foil or no foil), and perceived sweetness, speed of melting, texture (richness, thickness). Various feelings were associated with these sensory qualities (Figure 9.5). Consumers who tasted Hershey's Kisses said they were "sweet" and "quick melting." These qualities were found to be associated with overstimulation and joy. In addition, they perceived the benefits from Hershey's Kisses were to "get energy" and "satisfying a sugar craving." The Hershey's Bliss was described as "creamy," "rich" and "smooth." They melted at a slower rate than Hershey's Kisses. This formed the basis for a set of different cues associated with fulfillment of the concerns "contentment" and "stress relief."

These insights extended the design landscape. The slowness of chocolate melting and rich/creaminess of texture both became known as a key sensory cue for "contentment." The cue of a foil package was also associated with contentment. However, it was not clear from this research whether this is simply an association or a relevant relationship. The process of testing the impact of different design considerations, using this landscape, follows as the next step – the development of the package and packaged food design.

DEVELOPING THE PACKAGE DESIGN

In this book, the importance of a more holistic approach to development has been a continuous theme. This theme continues throughout the design process. The next few sections of the development of product design separate out design of

---

7. Generalized Procrustes Analysis, Society of Sensory Professionals, http://www.sensorysociety. org/ssp/wiki/Generalized_Procrustes_Analysis/. January 23, 2011

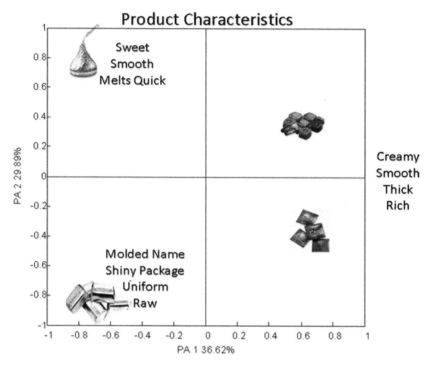

Product Characteristics

Creamy
Smooth
Thick
Rich

PA 2 29.89%

PA 1 36.62%

FIGURE 9.4 Free Association Mapping of chocolates on the basis of product characteristics.

packaging from the packaged food. However, this does not mean to imply that these two aspects of design should not occur concurrently, or that research should not include both package, and packaged foods. The intent is quite the contrary. These are separated solely to showcase their differences in approach to design.

The development iterations for the design of the package can be conceptualized into earlier and later iterations. In the earlier iterations, the focus is on understanding which cues contribute to the formation of anticipation emotions, through the design of front panel, and the formation of experience emotions during use. This is where, for example, the cue of foil or no foil would be determined as contributing to the anticipation of contentment, or the actual feeling of contentment through use. These early iterations give way to later stage learning designed to understand emotional reach – seeking to build a multiplicity of cues that reach out to form anticipation, and/or experience emotions through different cueing mechanisms. It may be that for some consumers, contentment cues arise from a foil wrapper or the color of the foil. For others, it might be the shape of the packaging containing the chocolate. The early iterations are focused on identifying cues that elicit emotions and which translate the promise of the concept into an experience. The later iterations are about understanding the market impact.

Feelings

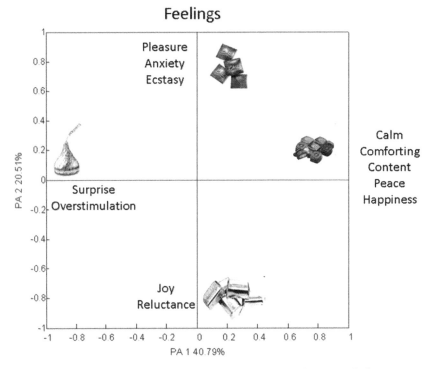

FIGURE 9.5   Free Association Mapping of chocolates on the basis of feelings.

Early Stage Package Design

All consumer goods have some version of packaging. Marc Gobe (2008[8]) describes its relationship to the actual product thus: "Packaging is a second half commercial." Packaging contributes to the overall product experience. It ensures quality by protecting the product between manufacturing and use. The merchandising of the product should create a positive display experience addressing where the packaging fits within the category, and differentiating the product with a recognizable shelf display.

The packaging experience helps to perform specific jobs-to-be-done (utility, social, sensory, psychological/emotional). For example, how a liquid laundry product pours was a revolutionary change in packaging. It is important to understand the utility issues of convenience, ease of use, and ergonomics. Packaging sensory factors need to address the overall aesthetics, any symbolic meaning, and should provide sensory cues about the product. The social aspect should convey the brand identity, while the psychological aspects should elicit

8.   M. Gobé, S. Zyman, Emotional Branding. Allworth Press, New York, NY, 2001

emotional and hedonic expectations. All together, the packaging should communicate, associate with, or elicit the brand's emotional qualities.

Early stage package design tends to involve small, highly collaborating teams of researcher, designer and consumer. These teams of co-designers tend to iterate through a process of "visual sense making," where designers listen and translate what they have heard into package protocepts of mock-ups – and consumers provide additional input.

Co-Design

The starting point for package design is the concept and collection of possible cues that come from the design landscape knowledge base. This starting point initiates an iterative dialogue between designers and consumers. The researcher contributes by listening and facilitating the translation between concept and package.

It is important that the team focuses on the relevant language of the consumer – as developed, refined and validated at the conceptual design phase. Further, it is important that the research makes clear distinctions between how elements of the conceptual design are intended to elicit expectations and what are the emotions from the anticipation of the experience (i.e. hope, low fear, intrigue, desire, and lack of disgust), or from use (i.e. experience emotions).

Packaging design can begin with simple paper protocepting as a creative, rough, perhaps hand-sketched drawing or wireframe of the package design. The concept is always shown first to keep a focus on what is being promised. Visual "sensemaking" and scenario building are design methodologies used to facilitate this iterative learning process.

As the design process continues, it may involve the iterative development of more elaborate package mock-ups, to better gauge emotional impact. This may include computer graphic creations of packaging labels or mock-ups that can be rapidly altered to guide the process of testing, listening, and dialoging.

This can be taken a step further using Computer Aided Design (CAD) protocepting to create a CAD model of design for assessment. These can be used to generate pictures or three-dimension representations of concepts for assessment, to look at pourable spouts, for example, or to rotate a graphic image of the product.

A more advanced technique is the capability to make solid free-form fabrication protocepting. A CAD drawing can be uploaded to a website and converted to an STL (stereolithographic) format, with images built in 3-D using a layer-by-layer construction that is built up of photosensitive layers representing each slice.

Mock-ups of label design can be printed on packaging "blanks" – cans and other complex packaging that can be used as early mock-ups. These are useful for experiential research where respondents, co-creators, and/or stakeholders can assess and provide feedback. They are also useful in concept development and prototyping label designs.

Later Stage Package Design

The early stage iterations conclude when the range of possible cues that might be built into the package design have been exhausted. At this point the question turns to how to optimize, achieving emotional reach across the consumer target. This requires the designer to get outside of the laboratory and into the field. At this point the design methodology becomes research-driven rather than design-driven. Below are a few techniques that might be applied to facilitate this process.

Choice-Based Conjoint Modeling

Howard Moskowitz and Alex Goffman, in their 2007 book *Selling Blue Elephants*,[9] recommend conducting conjoint modeling early in the design process to give designers better direction. In practice, conjoint modeling is a better technique in the later iterations – i.e. to help refine, not inspire creative package design. The design landscape serves as a much better way to inspire creativity through knowledge than the structure format of a conjoint analysis.

Choice-based conjoint was earlier introduced in Chapter 7 to identify the optimal combination of behavior drivers that have a common theme in defining the opportunity solution – and again in Chapter 8 to identify the optimal concept design. This technique is applicable to help refine and optimize packaging design, once the possible elements of design have been identified. It is a good method when there are a lot of prospective design elements to select, and a relatively few number of factors impacting the decision.

An example is a six-factor factorial design to find the combination of four design factors for a new packaging for squeezable margarine. There might be four different types of closures, four alternatives pictures, four package shapes, and four statements. It may be impossible to test all 164 combinations within a given study. Fractional factorial design and other more elaborate statistical designs enable the estimation, and testing of main effects, from choices among 16 different combinations of closures, pictures, shapes, and statements.

As in Chapter 8, the distinction here is made between choice-based conjoint and other conjoint methods such as ratings-based. Choice-base conjoint involves research participants appraising a series of package mock-ups or CAD drawings and selecting winners (or losers as with a "Max Diff" technique). The choices result in consumer quantitative information and statistical modeling to estimate the relative importance of specific elements to choice.

---

9. H.R. Moskowitz, A. Gofman, Selling Blue Elephants. Wharton School Publishing, Upper Saddle River, NJ, 2007, p. 145

Shelf Sets

In as much as competitive context is important to concept screening, it is also important to package design. This typically involves the testing of various package designs in shelf sets simulating the retail environment. Consider the case study for the Baked Lays packaging redesign, as reported by Frito-Lay and design firm Hornall Anderson.[10] Gannon Jones, vice president for marketing (Frito-Lay North America), states "Our efforts to connect with women are holistic ...".

In this article, Hornall Anderson describes their engagement with Frito-Lay to redesign the packaging for the Baked Lays product line. The consumer target is characterized as women looking for a healthier alternative, in the snack aisle. Their early design iterations involved creative development by evaluating "women-centric products" in other categories of the grocery store. Their focus was to understand what cues signal healthiness and appetite appeal. The article states that they then conducted qualitative research into how the protocepts (i.e. mock-ups) looked in shelf sets. The new package redesign is not shiny or metallic, but a more earthy – a natural color palette to signal healthy. Product photography was used to signal appetite appeal. The brand logo was made predominant, as shelf sets were found to contribute significantly to the appetite appeal of the product.

Shelf sets provide a more holistic perspective about how a package design stands out against the competition, within the context of a shopping experience. There are a number of different ways companies are testing shelf sets. One involves simulating the self set at a central location. Another involves the computerized simulation of shelf sets. Various software solutions exist[11] with a variety of applications.[12] A third is a mock-up of shopping experiences through the setting up of test stores.[13]

Choice Modeling

Research methods involve placing test packaging designs into various shelf sets with competitive products, which leads naturally to choice modeling. This type of research tends to be monadic, with research participants selecting packaging from one set to simulate shopping choice. As in selecting themes, filtering ideas, and finding the optimal combination of package design elements, choice

---

10. P. Lindsey, New Crisp Packaging Design from PepsiCo Targets Women, Bakery and Snacks. com, 3 April 2009. http://www.bakeryandsnacks.com/Processing-Packaging/New-crisp-packaging-design-from-PepsiCo-targets-women. January 18, 2011

11. FUSE homepage, Enhanced Survey Solutions, http://www.fuse.com.au/. February 26, 2011

12. W5 On Online Marketing Research, W5, http://www.w5insight.com/w5docs/whitepapers/W5%20on%20Online%20Marketing%20Research.pdf, 2007. February 20, 2011

13. Center for Advanced Retail Technology. http://www.bvinetworks.com/library/reports/ShopperGauge%20Case%20Study.pdf, 2011. January 21, 2011

is used to measure emotional impact. Different sets and/or test package designs can be incorporated to analyze which test design has greater emotional impact.[14] Experimental designs are typically used to vary shelf sets – not just the test packaging, but sometimes the location within the set. Statistical analyses tend to involve counting the frequency of choices, either against a benchmark or control package design.

However, as with choice-based conjoint modeling, choice alone will not always help in determining why a specific choice has been made. Therefore, additional diagnostic information is also often collected to help understand why elements of the design, in the context of competitive products, actually contributed to choice. By applying the appraisal framework, information about the underlying concerns and expectations for research participants can also be integrated into insights. In addition, interviews with research participants can be helpful to explain why choices are made.

DEVELOPING THE PACKAGE FOOD DESIGN

The development of the design for the packaged food should parallel that of the package design. This leads to an opportunity to understand how the package creates expectations, which impacts the packaged food design. The food designer needs to be someone capable of understanding how to design food that can be scaled up for commercial success. In the food industry, the food designer is the research chef. The research chef cannot operate independently; they must work closely with the innovation team, applying the same principles of interaction design as the package designer.

Similar to package design, food product design can be broken up into earlier and later iterations. The early iterations are to understand what cues might be built into the food product. The later iterations are to understand emotional reach.

The Research Chef

John Draz is a research chef working at Ed Miniat, Inc., a company providing meat products for food service and branded food product companies.[15] He prepares product presentations for customers, evaluates products, and works with product development to design food products. A research chef is a chef with expertise in not just the culinary arts, but also food science. Also called product development or food innovation chefs, research chefs are key members

14. J. Thomas, Choice Modeling for New Product Sales Forecasting, Market Research World. http://www.marketresearchworld.net/index.php?option=com_content&task=view&id=2929&Itemid=60, 2011. February 13, 2011

15. Research Chef Interview, http://www.myfootpath.com/advice-and-answers/career-interviews/research-chef-career-interview/. February 13, 2011

of the innovation team for food companies. They develop new foods and products for restaurant chains and food manufacturing companies.

As with packaging designers, research chefs utilize the principles of co-design to inspire and guide their creations. This process was described in an article about Anne Albertine, a research chef for Taco Bell (Crosby, 2002)[16]. In early testing sessions, Taco Bell brings in focus groups of consumers. These sessions start with concept research, where consumers choose among 50 or more pictures and written descriptions of possible menu items. Eventually, focus groups taste samples of the most appealing of the proposed foods. Responses are taken during experiments conducted in sensory labs by food scientists, sensory researchers, and marketers.

These examples of how research chefs operate highlight the iterative nature of how research chefs work with consumers and the innovation team to observe and learn. They help translate the promise of the concept into a product experience that delivers emotional impact.

Early Stage Packaged Food Design

The account of how Taco Bell's research chefs design foods for a food service company applies equally well to food product design for name brand food manufacturers. As with packaging design, the research chef is guided by the landscape of cues within the guardrails of the product concept. However, the research chef is challenged to design food solutions, within these guardrails, in other ways.[17] Not all creations in the lab can be easily converted into commercial products. Food designs must be able to be manufactured in food plants – versus a restaurant or commissary. Designed products must be able to be stored for extended periods – versus consumed within a few days. These food designs must also have food safety built into them. These challenges limit the food design to specific types of industrial ingredients. The research chef must not only design to cue emotional reactions among target consumers, but also to anticipate how the food might change as a result of its chemistry and manufacturing.

Interaction Food Product Design

It is very important for the research chef and package designer to work together to overcome the challenges from the interactions among package, food, and consumer. The side panels of packaging can provide information to consumers through its ingredient statements. Key words on ingredient labels may have

16. Olivia Crosby, You're a *what*? Research chef, Occupational Outlook Quarterly Online, Fall 2002, vol. 46, Number 3, http://www.bls.gov/opub/ooq/2002/fall/yawhat.htm. January 10, 2011

17. K. Ware, Food Science Principle for the Chef, Presented at Institute of Food Technology Annual Meeting, 16–20 July, 2005, New Orleans, LA

different meaning to consumers, leading to the formation of various expectations about the food experience. Package windows can provide visual display of the package food contents inside. These can lead to cues that alter consumer expectations about what to expect from the food consumption experience.

The package design can also lead to functionality in the preparation of the food for consumption. This includes closures, hand grips of value to facilitate opening and using packages. Microwave cooking technology requires the package to not only hold the food, but also to help in cooking the food. This interaction not only impacts how the food is formulated, but also how the consumer cooks the product.

The presence of these interactions complicates the design considerations for both research chef and package designer. When package mock-ups are available, they can be of great help to research chefs. Likewise, the design of food can impact the requirements of the package designer. For this reason it is often helpful for the research chef and package designer to work together in the early design stages.

In working together, there are shared limitations in how mock-ups and protocept food creations can be assessed by research participants. Package mock-ups tend to be expensive to build – limiting the number that can be easily developed for feedback from consumers. The research chefs tend to cook up food creations in small batches, also limiting the number of consumers that can provide feedback. For this reason, a highly collaborative co-design process is typically used to bring consumer feedback to both the research chef and package designer. This approach to design involves engaging small groups of research participants to assess the product concept, package mockup, and chef-designed food creations.

Holistic Research Methods

These early stage iterations of food product design lend themselves well to holistic research methodology. Consider the case study of a beverage manufacturer interested in optimizing the colors of a new line of flavored beverages. In this case, the line concept had been selected at the front end of innovation – the discovery phase. The product names were developed using a tournament and TURF analysis. The beverage was consumed in a clear bottle. Therefore, there were complex interactions to consider between the name of the product, and the flavor and color of the beverage.

The moments-of-truth design, introduced in Chapter 5, was applied in this research study – in order to increase learning to select the best flavors and to optimize the color for each of three products in the line. Research participants were each shown the product concept – including the product name – and then asked to comment about the packaged product. Each packaged product (fully labeled) was then introduced in its three alternative colors. Participants selected the color best fitting the product concept and product name. This colored beverage was then poured into a glass for consumption and appraisal.

The research design provided valuable information about the interaction between concept, color and taste experience. In addition, each participant was asked to appraise a shelf set including three different products (with one of the respective test colorings for each shelf set). The final analysis served as a confirmation of how well the product color not only communicated the product name (i.e. the flavor), it also served to help assess the colors against the competition.

In this way, the more holistic the early iterations of food product design, the faster the iterative learning about the complexities of the food consumption experience. For this reason, research chefs – such as Anne Albertine and John Draz – must work closely with other members of the innovation team. Their food designs must not only overcome the challenges of the research chef, but take advantage of underlying interactions that contribute to the success of food products.

Later Stage Design Iterations

We have, so far, characterized food product design as a process by which innovation teams develop product requirements defining elements of the package and packaged food to hit the target – the experience envisioned from the Discovery phase as the best opportunity. The product concept is the guardrails for the package design, within the Design landscape. So too, must the concept be the guardrails for the packaged food design. However, it is at this point in the product development process where the focus on the target is sometimes lost. The result is a food experience that does not fully live up to the promise of the concept.

This loss of focus sometimes occurs during the transition from research chef to product developer. This is the point where the process of food product design becomes research, rather than design, led. The focus shifts away from hitting the target as promised by the concept, to making the food more pleasurable, to maximize "liking."

As mentioned earlier in this chapter, Liking is a very generalized, non-descript response measure. Its projection as an emotion can vary, from directed to the product itself to specific features of sensory qualities. This focus is often the result of how research is designed, rather than a natural response to the product experienced as a whole.

If prior phases of product development discover that the differentiating quality of a product is its promise of freshness, then the reliance on Liking as a measure of pleasure can sidetrack the research chef and product development. In the absence of any better metric, Liking is better than nothing. However, there are often better guidance metrics when the goal is to translate the concept into a food product design. For this reason, one of the more important objectives for packaged food design is to discover the consumer metrics upon which the product development team can rely to guide their development activities.

Rapid Iterative Holistic Design

It is possible to apply the moments of truth, and other more holistic research designs, into a rapid iterative design process where package and packaged product are designed together. This involves design, testing, learning, and redesign. It also requires a high degree of participation from the innovation team.

An example is the case study to develop a new breakfast beverage. This case study involves the application of a rapid iterative holistic design over a five-day period. A number of alternative beverage formulations had been identified as rough protocepts, based upon landscape research into what might be considered a "premium" product by target consumers. Concept research had resulted in the identification of the language of the consumer with regard to what is considered premium in this category. "Freshness" was one such attribute, with various associated sensory qualities of mouth feel, flavor, and taste.

The rapid iterative test methodology involved cycling between days in the field at a central location, dialoging with consumers to gain new direction, and then a day in the laboratory to reformulate. The research made three such iterations over five days: day 1 in field, day 2 in lab, day 3 in field, day 4 back in the lab, and day 5 back in field. Within each day, there were five sessions of 40 research participants. The protocepts were compared to freshness and other consumer language associated with premium. Innovation team members were able to see research results and then to select specific research participants for one-on-one interviews. The learning between each session, at the central location, resulted in rapid understanding for what was considered premium. Not only was freshness better understood, but the rapid iterative learning led to deeper understanding of how different consumers translate premium into expected experience. This rapid, adaptive approach to design not only resulted in the rapid design of a beverage prototype, it resulted in the development of a product that has exceeded all expectations for the brand in the marketplace.

DESIGN VALIDATION

There are two types of validation to confirm that a design has achieved its goals: one is internal and the other external validation. Internal validation is similar to the validation methodology used at the end of the concept design phase. This involves the testing of the concept against a benchmark product experience. The concept is validated by testing whether the benchmark experience is impacted by introducing the concept before (versus after) the benchmark experience.

Internal validation involves testing whether the experience confirms, meets, or disconfirms the expected experience as promised by the concept. The language used in the concept statement is confirmed for the package design and the packaged food design. Internal and external validation can also be conducted in the same research study – by comparing how well the benchmark achieves internal validation against the test product design.

Key Points

- Food product design is a translation of the expectations generated by a concept into a product experience.
- The goals of food product design are to change behavior through anticipation emotions elicited by package design and experience emotions through packaging and packaged food.
- Design for emotional impact involves the building of cues into packaging and packaged foods.
- Interaction design involves holistic considerations into how the concept, package, packaged food, environment and consumer interact to form experiences.
- Design methodology can be characterized on the basis of the use of inductive or deductive reasoning or taking a more positivist or interpretive approach to learning through research.
- Food product design requires a dialogic design process involving small groups of consumers engaged in dialogue with food product designers.
- Package and packaged food creative design is based upon the food product design landscape. This landscape is more granular than the concept landscape, built from consumer product insights into the design elements and intrinsic qualities that cue emotions through food product use and consumption.
- Free association profiling is a methodology to understand the food product design landscape, through the generation of associations between food perceptions, emotions, imagery, and benefits.
- Early stage package design tends to involve many rapid iterations, using dialogic co-design methods led by the package designer.
- Later stage package design tends to involve fewer consumer research studies, applying choice-based conjoint modeling and choice modeling of shelf sets.
- The research chef is someone trained in the culinary arts and food science. The research chef leads the design of packaged foods, using more interaction design methodologies in collaboration with small groups of consumers.
- Later stage food product design tends to involve holistic methods and rapid iterative research to design both package and packaged foods.
- The validation of food product design involves both internal and external validation. Internal validation involves the testing of food product harmony between the promise of the concept and the experiences. External validation involves testing the impact of a new experience on habitual consumption and use of an incumbent.

Chapter 10

Development

> *A person buying ordinary products in a supermarket is in touch with his deepest emotions.*
>
> John Kenneth Galbraith

HOLISTIC PRODUCT DEVELOPMENT

The development phase involves facing some of the most complex challenges within the Product Development Cycle. In this phase, the product specification is built from requirements – not just experience requirements, but a set of challenging commercial requirements. This specification starts with a working prototype that must be refined into a gold standard, and finally developed into a commercially viable product. This must be done without compromising the experience it must deliver.

To achieve success, developers have to factor in the complexities involved in commercialization. This requires an innovation team effort. Product development – guided by sensory research – cannot be successful developing in a silo. While product developers have leadership roles in this phase, they require team support to ensure the product hits the experience target. All must help the brand take the target consumers to the destination they seek.

Commercialization requires technical knowledge to overcome the many challenges involved in commercial development. Product developers include a diverse group of domain experts – food chemists, packaging engineers, food engineers, and food microbiologists. These members of the innovation team are typically the most scientifically minded, with a tendency to apply scientific methods to extend their knowledge. However, the technical complexities of commercialization require that they apply a more holistic perspective, throughout the development phase, to ensure success.

This holistic perspective is maintained with the help of the sensory researcher. The sensory researcher is the champion on the team for the consumer experience – keeping product development focused on the experience, rather than the product. This focus will help ensure that the experience requirements are not traded away, as development strives to factor commercial requirements into their development objectives.

Breakthrough Food Product Innovation Through Emotions Research. DOI: 10.1016/B978-0-12-387712-3.00010-4

In this chapter, a number of research methods will be discussed to keep the focus on the experience. An expanded sensory toolbox will be introduced to deliver fast, actionable consumer product guidance – to build a commercial product optimized for emotional impact within contexts of use and consumption.

The outcome of the holistic product development phase is a final specification for the delivery of the product to the marketplace. It involves a final validation that a fully commercialized product can, in fact, hit the experience target – achieving the experience impact desired by consumers, and achieving the market impact envisioned by the innovation team for the brand.

DEVELOPMENT GOALS

Development is simplified by a good design. The design provides a clear focus for what is the target, such that the skills of product developers can be applied to overcome the challenge of providing the experience to the marketplace. Design is the planning process that lays the basis for the making of an object.[1] As a verb, "to design" refers to the process of originating and developing a plan for a product. In contrast, the verb "to develop" refers to the process of building the product. Anil Mital, an academic leader in product development process theory, makes a clear distinction in his book *Product Development*.[2] Mital and his co-authors state that "thought must be given to how easily or not it [the product] will be manufactured, how difficult or not it will be to assemble and disassemble for the end-use-customer and how much it will cost in materials, production and maintenance."

The overall goal of food product development is to develop a specification for each product in a given product line. This specification is the complete set of instructions defining the products, not just in terms of package food qualities and package design elements, but also in terms of sourced materials for the building of these product components. The specification also defines how the product will be manufactured and controlled, to ensure their quality not just in manufacturing, but in shelf-life stability from manufacture to consumption.

This specification also must consider a number of guardrails. These guardrails not only include considerations of importance from earlier phases of product development, but also new considerations such as regulation, the environmental conditions for food product warehousing, distribution and merchandizing, and cost.

1. Design, http://en.wikipedia.org/wiki/Design. January 29, 2011

2. A. Mital, A. Desai, A. Subramanian, A. Mital, Product Development: A Structured Approach to Consumer Product Development, Design, and Manufacture. Butterworth–Heinemann, Oxford, 2008

Guardrails

The guardrails continue to narrow the scope of the product definition within which to develop a product specification. As the focus narrows on what is to be developed, more knowledge must be gained about the granularity of the product landscape within those guardrails. There are several different guardrails within which to develop the product. This includes a definition of who the target consumer is and any technical, or market-driven, platforms that are determined from the innovation strategy. It also includes the experience theme from the discovery phase. The concept elements from concept design and the experience requirements from product design also contribute to form guardrails.

Experience Requirements

The requirements coming from the design pertain to how the experience is made manifest by the product. To be fully behavior-driven, these requirements include the following considerations:

- Performance: how well the product experience delivers on its promised jobs-to-be-done (functional, sensory, social, psychological)
- Features: how differentiating the performance qualities of the product are as experienced against the competition
- Aesthetics: how the product sensory experience achieves emotional impact on a consumer
- Reputation: how the product experience will impact future assessments of the product and brand by communities

Commercial Requirements

Commercial requirements include technical requirements, as well as financial requirements with regard to the cost of manufacturing, distributing, and merchandizing the product. The technical requirements for commercialization include the following considerations:

- Translation: how well the product requirements manifest the product design
- Conformance: how well the product meets regulatory, industry standards, and standards of identity
- Reliability: how consistently the product delivers the experience over time (i.e. its manufactured quality)
- Durability: how long the product maintains an acceptable standard of performance (i.e. its shelf-life stability)
- Scalability: how the design can be replicated for commercial manufacturing (i.e. its materials sourcing and industrial processing)
- Distribution: how the product can be distributed to retail without compromising performance

- Disposal: how components of the product (e.g. packaging) can be disposed of or recycled
- Merchandising: how the product will be marketed and displayed in retail stores

For example, consider the development of an organic, instant coffee. There may be a market opportunity for this product amongst an identified target group of consumers. The concept might test well, suggesting target consumers would seek out and choose this product if it were available. The packaging and packaged instant coffee might also be designed, leading to a set of experience requirements. However, there might be an issue in sourcing a sufficiently large supply of organic coffee to support envisioned distribution to the marketplace. Further, there might be technical constraints in the types of ingredients used to meet the standard of identity for an instant organic coffee.

Commercial constraints such as these can sufficiently limit the economical feasibility of a set of experience requirements. Therefore, it is important that the innovation team maintain a very holistic perspective when considering how to overcome these challenges. The collective intelligence of the whole team is often the difference between success and failure in delivering upon the promise of the experience.

APPROACHES TO DEVELOPMENT

Quality Functional Deployment (QFD)

There are many approaches to product development in the food industry. One common holistic approach to product development strategy is that of Quality Functional Deployment (QFD).[3] Co-developed by Japanese quality industrial designer Yoji Akao[4] in 1966, this methodology has been refined and extended into a four-step process including product design, development, manufacturing, and quality monitoring.[5]

The value of QFD is in how it provides a thought process for how to take the experience requirements from the design phase to address the challenges in commercializing – i.e. taking into account commercial requirements. QFD can be incorporated into behavior-driven innovation as three stages of product development: (1) product optimization; (2) process optimization; and (3) process/quality control. Each of these three stages results in a different set of technical specifications for the commercialization of products.

---

3. Quality Function Deployment, http://en.wikipedia.org/wiki/Quality_function_deployment. January 12, 2011

4. Who Is Dr. Akao, QFD Institute, http://www.qfdi.org/what_is_qfd/who_is_dr_akao.htm. January 12, 2011

5. Customer-focused Development with QFD, http://www.npd-solutions.com/qfd.html. January 12, 2011

In product optimization, the gold standard from product design is refined by taking into account the diversity in cues and preferences that exist within the market of target consumers. Further, complex interactions between and among the packaging components and packaged food ingredients must be considered to maintain the product experience. The results of this development stage are specifications for the sourcing of alternative food ingredients, and packaging materials to ensure that these different components can be interchanged through procurement without impacting the experience. This also includes specifications that ensure these components and the product, as a whole, conforms to industry regulations and standards.

Overcoming these challenges often requires inspiration and guidance through sensory research. Specific sensory methods have been developed to ensure that challenges associated with conformance, component tradeoffs, and other product component interactions are overcome to keep the product experience on target. The behavior-driven approach requires the application of a more broad set of tools to understand and account for cue and preference diversity. These will be discussed later in this chapter.

Process optimization involves scaling the production process from pilot plant to a full commercial manufacturing operation. It is important to note that food processes rarely scale up "linearly." Scaling processes require technical knowledge into the complexities of food chemistry and engineering. Sensory research can play a chief role in helping process engineers overcome these challenges, by keeping the focus on the delivery of the target experience. This is done through research resulting in the building of predictive models interrelating technical product development measures to sensory measures.

Knowledge as to how a product will be industrially manufactured provides the basis for the final stage of QFD. This stage is process/quality control. It includes the development of specifications to monitor manufacturing and to ensure continuous quality of the manufactured product. This also includes quality considerations associated with the reliability, durability, distribution and merchandizing of food products.

Reliability involves the development of quality specifications associated with the manufacturing process itself – assuring that the manufacturing process is monitored to keep it in control. This typically involves developing a wide range of specifications for testing sourced materials, manufactured partial product, and final product. These often involve a mix of physical, chemical and sensory specifications, as well as process control measures (e.g. rates and temperatures).

Product durability results from specifications associated with the shelf-stability of the product. This involves complex interactions between and among the packaging, and the packaged food components in different environments. Knowledge is required as to the factors that stress products and shorten shelf-life. Knowledge is also required regarding the environment conditions post-manufacturing, including warehousing, distribution, merchandising, and in-home storage prior to usage and consumption. Durability requirements can be achieved by applying this knowledge to consider solutions that require change in packaging materials, formulation, and even processing.

Sensory research is an essential contributor to the building of this knowledge. It also has a role to play in testing various development considerations, and to overcome challenges to provide the designed experience in a commercial food product that is reliable and durable throughout its distribution, merchandising, and in-home storage. The resulting specifications tend to include in-plant manufacturing control targets and tolerances that define quality. However, quality is ultimately defined by the consumer in how they experience the product through use and consumption. The sensory researcher must help in defining quality according to the consumer. This is typically done by interrelating consumer to sensory, chemical, physical, and instrumental measures, such that control targets and tolerances can be specified.

The level of complexity in bringing in the voice of the consumer increases in a global manufacturing environment. Many of these issues go beyond the scope of this book, yet are important to be understood to ensure that the actual consumer experience is consistent with the experience in the food product design phase.

Development Impact on Consumer Behavior

While concept design is important in motivating selection (trial) and searching behavior, development leads to product experiences that motivate repeat trial and sharing behavior. Although the design phase makes this a much easier task for innovation teams, development provides the experience promised by the concept. For this reason, the development phase directly impacts innovation success.

Further, the product optimization stage of development is the part of the process where key decisions in the delivery of the envisioned experience ultimately succeed or fail. Decisions based upon constraints (e.g. costs, regulations, supply of ingredients) often lead to tradeoffs. As a result, not all design qualities and elements make it into the final product. These decisions can severely impact the ability of innovation teams to achieve their consumer behavior goals. Therefore, innovation teams must maintain a holistic approach to development.

It is also at the product optimization stage of development where broader, market level input from target consumers results in knowledge on how the product experience leads to emotional reach. In the design phase, prototype amounts are typically limited, constraining how many research participants can provide feedback. In the development phase, prototypes are more rapidly produced in a pilot plant, in sufficient quantities, to field large-scale consumer research throughout the target market.

Once the formulation and packaging components are set, the development focus turns to developing specifications for process optimization, as well as process and quality control. This focus is important to ensure product durability, scalability, and reliability, yet is less important to the overall success of the product. Durability and reliability are consumer requirements, not market differentiators. In most cases, if a product can be produced in a pilot plant, it can be mass manufactured. In most cases, process and quality control will not greatly constrain the optimized product experience. Therefore, the remainder of this chapter will be focused primarily on the first stage – product optimization.

Sensory Hurdles

Sensory hurdles are consumer response measures that have meaning in guiding product development. They provide a metric to gauge progress toward development goals. There are four types of sensory hurdles: (1) technical sensory statistics; (2) measured consumer responses; (3) probability for a measured consumer response; and (4) probability for a consumer behavior.

Technical sensory hurdles are the experience requirements from the design phase. They are the differentiating sensory descriptive qualities of the gold standard, versus the competition. Sensory descriptive analysis can be applied to profile, and characterize what sensory descriptors differ significantly between a gold standard and the competition. The measured values of these differentiating qualities can serve as hurdles.

Often, there is not sufficient knowledge about whether or not a differentiating sensory quality is a driver of a consumer response. In these cases, consumer research is necessary to build this knowledge. This is where the other three types of sensory hurdles can be applied to help guide development.

Consumer responses can be defined as a statistical "measured response" to a given product by a group of research participants. An example of this is a measured response to Liking on a 9-point hedonic scale. By averaging responses to a number of products by a group of target consumers, a distribution of product scores can be used to establish a hurdle. Typically, hurdles such as this are based upon percentiles. For example, the 80% percentile of mean Liking scores from target consumers to a group of "successful" products might serve as a hurdle. In the case study of Baked Lays (Chapter 2), the product was reported to be released to market after achieving a 7.2 score in Liking.

Sensory hurdles can also be defined as the "probability for a measured response." Consider again the research participants who rated the Baked Lays product on Liking (9-point scale). Instead of using an average as the sensory hurdle, the percentage of consumers scoring the product as 7 or above can be used as an estimate of a "measured response." This measured response takes on more meaning than a simple average when it can be associated to motivate behavior.

The finding of a single – univariate – consumer response measure that predicts behavior is unlikely in most cases without assumptions. The more likely case is to develop a predictive model of behavior as a function of several – multivariate – consumer response measures. If developed, this would be an example of a "probability for a behavior" hurdle – e.g. what is the chance that an experience with a product will result in a given type of behavior. The appraisal framework provides a basis to establish these types of hurdles. This also requires the "training" of models through historical data. By collecting the same set of consumer response measures, a model can be built to calculate the chance that any randomly chosen cue will exhibit the response behavior.

When the goals for the innovation team have been defined in behavior terms, then these four types of hurdles take on a different significance. Consider the example where the goal is to replace habitual use of an incumbent product with a new product. Hurdles based upon technical sensory statistics require the application of a heuristic model to infer that the sensory defined hurdle (e.g. sweetness level) results in this behavior. This inference (sweetness hurdle $\rightarrow$ behavior) requires a lot of assumptions. Hurdles based upon a measured consumer response (e.g. 7.2 in Liking), or probability for a measured consumer response (e.g. 75% scoring a 7 or higher in Liking), requires increasingly fewer assumptions. Hurdles based upon a probability for a consumer behavior simply require a good behavior model with model inputs that can be generated from consumer response information.

The "Gold Standard" as an Experience Target

As has been noted earlier in this chapter, the sensory researcher must become the champion for the consumer experience. A gold standard is defined in this book as a food prototype and package mock-up that serves to recreate the experience among target consumers. It is an essential tool that the sensory researcher can use to maintain focus on the experience throughout development. The ability to rapidly and accurately generate this gold standard is a huge advantage for an innovation team. This not only helps to speed up the product development process, but it also helps ensure the specification hits the target.

Without the design phase, product development is often left without a gold standard. By associating the gold standard formula and elements of design to the experience, a roadmap can be built for development to know how to deliver the experience promised by the concept. Direct translation from concept to development without a gold standard may work for simple line extensions, but rarely for new – breakthrough – food products. The knowledge gap is simply too big for product developers.

Sensory research plays a key role in ensuring that the sensory aspects of the target experience can be recreated through a gold standard. This involves the use of descriptive analysis to provide a technical sensory profile of the gold standard. There are several different methods for descriptive analysis[6,7] – all involving trained human subjects to characterize and profile the sensory properties of foods.

The descriptive profile of the packaged "gold standard" food, and associated experience requirements, typically provides sufficient information to build a gold standard specification. This specification typically ensures that the gold standard can be rapidly produced in sufficient quantity to support consumer product research. Quantities are required to generate statistically significant feedback from target consumers, acting as research participants from a sample of markets. Production is typically done in a pilot plant where standard and experimental processing equipment can be used for test productions. Packaging typically does not require sensory profiling to be recreated for consumer research – relying instead on specifications based on standardized elements of form, artwork, and content.

Consumer Tolerances

The concept of consumer tolerance is vitally important to the behavioral approach to product development. Tolerance is defined as the degree of deviation from the gold standard (as defined by descriptive analysis and packaging specifications) that is "tolerated" by the consumer – i.e. does not change the experience. Consumer tolerances are important in the building of specifications. While a gold standard leads to the definition of a product target, the consumer tolerances form the basis for specification ranges.

Consumer tolerances vary according to the degree of difference from a gold standard. Tolerances can be defined as no perceptual difference, no recognizable difference, or no difference in preference. From a behavior perspective, the choice of tolerance is important in determining what research methods are most appropriate to guide product development.

Tolerances tend to be fairly wide prior to the launch of a product into the marketplace. Once a product has been released, the tolerance rapidly decreases. The overarching determinant of tolerance depends on what degree of sensory change results in either a decrease in positive emotional impact, or a decrease in fulfilling the promise of the experience (i.e. degrading the translation of concept into product experience).

Prior to launch, it is important that the product should simply deliver on the promise of the experience through packaging and marketing communications. In this way, the gold standard from the design phase may have

6. H. Stone, J. Sidel, Sensory Evaluation Practices, third ed., Elsevier, Amsterdam, 2004

7. M. Mailgaard, G.V. Civille, B.T. Carr, Sensory Evaluation Techniques, third ed., CRC Press, Boca Raton, FL, 1999

a fairly wide tolerance. This is good, as there may need to be considerable refinements to the final product specifications in order to overcome the many product development challenges. Sensory attributes that serve as consumer standards of identity are also important aspects in determining tolerance. Consider the standard of identity for a product that is to fit into the beverage cola category. In this category, there is a fairly well-defined "schema" (as defined in Chapter 9) for what is and is not a beverage cola. The category of product experiences forms a tolerance around the degree of carbonation, degree of sweetness, and sweet–sour balance to form a schema in the minds of consumers.

When a product is so new that it defines its own category, then consumer tolerance is greatest. An example of this is Red Bull which defined the energy drink market. Created in 1987, by Austrian entrepreneur Dietrich Mateschitza, Red Bull was released into Europe in 1989 and finally into the United States in 1997.[8] Today, it is the market leader among energy drinks. However, it did not receive high positive scores when it was tested in a blind sip test (Wipperfürth, 2003).[9]

These expectations become more set once a consumer repeats the experience. Further, and more importantly, sensory attributes that cue emotional impact during a product experience tend to require a much smaller consumer tolerance. The unpleasant flavor of the original Red Bull product offers a unique set of sensory cues that elicit positive emotional impact through use. However, if product use is to be habitual, it is important that there is consistency in use, to avoid awakening the conscious mind with the awareness that "something is different" about the product experience.

SENSORY RESEARCH

Sensory research is essential to generate the knowledge necessary for the innovation team to develop food products. As discussed in Chapter 3, the sensory evaluation field evolved out of the laboratory in using frameworks as a basis in perceptual psychology to develop applications. Today, sensory applies its toolbox of applications to generate consumer product insights in support of these three stages of food product development: product optimization, process optimization, and process/quality control.

In each of these areas of application, the sensory researcher has developed methods where insights are generated, by interrelating consumer information and technical product development measures. Technical product development measures can be discrimination sensory measures generated by trained

---

8. Red Bull, Wikipedia, http://en.wikipedia.org/wiki/Red_Bull. January 25, 2011

9. A. Wipperfürth, Speed-in-a-Can: The Red Bull Story, *Plan B*, 2003 http://www.slingstone.us/uploads/Speed_In_a_Can.pdf. July 6, 2011

panelists, using a quantitative and/or qualitative approach to evaluations in the sensory laboratory or at in-plant quality control laboratories. They can also be physical, chemical or instrumental measures generated from laboratory or processing equipment that has been shown to predict consumer response or discrimination sensory measures.

Consumer information is often more expensive and slower to generate than sensory discrimination measures. Typically, non-sensory technical product development measures are the least expensive and quickest to attain. Therefore, one role of sensory research is to develop predictive models of consumer response behavior that are based upon less expensive and faster generating technical product development measures.

In this way, sensory research involves methods that result in the building and use of statistical and heuristic models. There are three ways in which models guide development: (1) identification of less expensive technical measures that can be used to guide development towards the target experience; (2) establishment of hurdles and consumer tolerances; and (3) formation of specifications that define the boundaries for manufacturing a product that delivers the target experience.

Product Guidance Methods

Most sensory research teams that support product development in the food industry tend to think of themselves as providing product guidance research. To guide product development, their toolbox not only includes consumer research, but also laboratory measures of sensory perception. The latter tools involve the training of panelists to perform sensory difference tests, quality assessments, or descriptive analyses. They also utilize some qualitative methods.

Figure 10.1 summarizes what is in the sensory toolbox. These methods and techniques can be characterized in what they measure (i.e. sensory affect or sensory perceptions), how they are measured (i.e. qualitative or quantitative), and in what context they are measured (i.e. laboratory, central location, in-home/in-context use).

Holistic Research methods are clustered in the upper half of the schematic. Most involve a hybrid of quantitative and qualitative methods. Sensory discrimination methods, involving trained panels to characterize sensory perceptions, are clustered in the lower half of the figure. In the lower right are the quantitative methods for Descriptive Analysis, and the respective Perceptual Mapping techniques (e.g. principal components analysis) to characterize how products differ in their sensory perceptions. In the lower left are the more qualitative descriptive analysis methods, including Projective Mapping,[10]

---

10. E. Risvik, J. McEwan, M. Rodbotter, Evaluation of Sensory Profiling and Projective Mapping Data, Journal of Food Quality and Preference 8 (1) (1997) 63–71

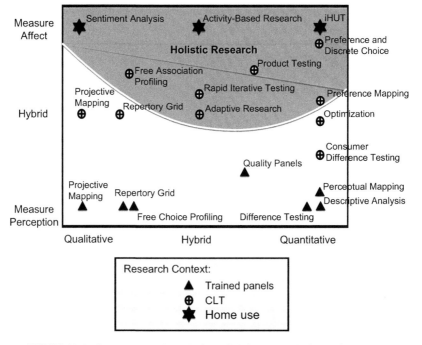

FIGURE 10.1 Sensory research methods applied throughout the innovation process.

Repertory Grid,[11] and Free Choice Profiling.[12] Quality Panels involve trained judges who measure a food product according to specifications for process and quality control.

The more controlled central location testing (CLT) methods, involving the recruitment of consumers as research participants, are clustered in the middle range of the grid. This includes the purely quantitative methods of Preference and Discrete Choice Testing, Preference Mapping, Optimization, and Consumer Difference Testing. It also includes qualitative methods and hybrid methods of Projective Mapping, Repertory Grid, and Free Association Profiling using consumers. Product Testing, Rapid Iterative Testing, and Adaptive Research methods tend to include information about consumer response behaviors (e.g. affect), as well as other information to help understand why affect is elicited by product experiences. These methods tend to be the most holistic, comprising of hybrids of both affect and perceptual information, and of quantitative and qualitative measures.

11. J.A. McEwan, J.S. Colwill, D.M.H. Thomson, The Application of Two Free Choice Profiling Methods to Investigate the Sensory Characteristics of Chocolate, Journal of Sensory Studies 3 (1989) 271–286

12. A.A. Williams, S.P. Langron, The Use of Free-choice Profiling for the Evaluation of Commercial Ports, Journal of the Science of Food and Agriculture 35 (5) (1984) 558–568

At the top of Figure 10.1 are methods that are focused mostly on measuring consumer product affect in the context of experiences. In the upper right is the quantitative in-home use test (iHUT). In the upper left, is sentiment analysis as applied to product testing with consumers (Social Dialogue, Chapter 7). In the upper center is Activity-Based Research, which is a hybrid of quantitative and qualitative research.

This toolbox of methods is applied to generate knowledge about the product landscapes, within which developers are inspired to create and guided to make decisions. It is the hybrid techniques that generate the most diverse mix of data – and, as a result, can lead to the greatest opportunity to integrate diverse data into insights. It is the collection of Holistic Research Methods, at the top of this figure, that are changing the face of product development.

Consumer Response Measures

Sensory research has traditionally relied on a fairly small set of consumer response measures. The behavior-driven approach to product guidance focuses on the following consumer response measures to guide development: difference judgments, choice, self-reports on emotions, qualitative storytelling and commenting, and behavior observations. There are a number of other response measures that have been traditionally used by sensory researchers, as diagnostic measures, to guide development. This includes the response measures of just-about-right (JAR) and consumer self-reports of perceptual descriptors.

Difference judgments are used more frequently with trained panels than with consumers. Their respective questions and measures are as follows:

- Pure Difference – Are these products the same or different ("Same–Difference Test")?
- Difference from Control – How different are these products ("rated on a scale of difference")?

There are a limited number of situations where difference judgments, by consumers, can help guide development. This includes situations where a difference judgment is an appropriate measure of consumer tolerance. Such can be the case with habitually used products of great importance to the consumer. In these cases, any perceived deviation from expected is subject to a strong negative emotional reaction. However, consumer difference judgments should be used cautiously. Care must be taken in the design of difference tests to ensure that detected differences are those of normal sample variation. This can be managed by taking a more behavioral approach to consider how a product is consumed, what context it is normally consumed within, and what portion sizes contribute to a meaningful judgment.

Liking is the most common consumer response measure generated to support product development in the food industry. A typical consumer response measure for Liking is generated as follows:

- Liked – Please rate how much you liked or disliked this product (rating on a balanced line or category scale).

There is a general belief among many sensory researchers that Liking is purely a response to perceived sensory pleasure. In an atomistic – highly controlled – experiment this might be accurate. However, this definition no longer holds when research becomes holistic. Liking in a holistic context is an emotional response projected onto the product. The eliciting condition for this emotion is defined as "Liking an appealing object" (see Table 4.1).

It should be noted from the Emotions Insight Wheel that Liking can occur within any of the four quadrants. You can "like" (or "love") a product for its functionality, its sensory pleasure, its self-social identity meaning, or its meaning in changing your emotional state of mind. Therefore, in a holistic research context, Liking is a general, non-descript measure of emotional impact.

There are better response measures than Liking to gauge how a product impacts an experience. These response measures go beyond Liking by gauging different types of emotional impacts from the experience. Among the experience emotions defined in the emotions topology in Chapter 4, there are six that arise strictly from eliciting conditions projected onto the experience (rather than the product). These response measures gauge how much the experience was enjoyed: a relief, disappointing, satisfying or dissatisfying, surprising, and/or boring. The following questions can be used to generate these as self-reported measures, immediately following product appraisals.

- Enjoyed – Please rate how much you enjoyed this experience (rating from not at all to enjoyed extremely)
- Disappointed – Please rate how disappointed (if at all) you were with this experience (rating from not at all to extremely disappointed)
- Satisfied – Please rate how satisfied or dissatisfied you were with this experience (rating on a balanced line or category scales)
- Relieved – Please rate how relieved you were (if at all) by this experience (rating from not at all to extremely relieved)
- Surprised – Please rate how pleasantly or unpleasantly surprised you were with this experience (rating on a balanced line or category scale)
- Boring – Please rate how boring (if at all) was this experience (not at all to extremely boring)

Perhaps the most simple consumer response measure is to ask consumers to make a choice. This method was shown to work well in concept research, as an emotions-based metric when forcing research participants into making many quick choices (Chapter 8). When there is no time to rationalize, research participants are forced to rely on their emotions to choose. However, this is not the case in product appraisals that involve consuming and using products. Development phase product appraisals are typically limited to a smaller set of appraised foods, due to a number of psychological and physiological problems associated with too many product assessments.[13]

Choice measures can include a gold standard or reference in the choice set. When there are two alternatives, the method collapses to a typical preference test. Otherwise, the method expands to discrete choice testing as discussed in Chapter 9.

- Preference – Which product do you prefer (forced choice, or a no preference and/or a both equally preferred option can be added)?
- Choice – Which of the products did you most (least) prefer ("Discrete Choice" or "Maximum-Difference")?

It is often too difficult a task for research participants to rate the degree of how much a product makes them feel. There are five experience emotions projected onto oneself or onto others: shame, jealousy, contempt, admiration, or pride. When these emotions are relevant to product development, it makes sense to simplify the judgments into a simple Check-All-That-Apply list of measures.

- Check-All-That-Apply – Which of the following list of attributes applies to how this product made you feel (presence/ absence of the emotions descriptors of shame, jealousy, contempt, admiration, or pride)?

The final experience emotion in our topology is anger. Development goals can sometimes lead to a need to measure anger. This includes cases where there is a need to solve a problem by minimizing anger, or when messaging is so disruptive that the new concept and product experience leads to increased anger against a competitor.

- Angry – Please rate how angry you are (if at all) at the [brand/manufacturer] for doing this to you (rate not at all to extremely angry)

There are five basic anticipation emotions (fear, hope, intrigue, desire, and disgust). Fear, hope, and intrigue are purely associated with the anticipation of first time trials. However, desire and disgust can equally apply well to the anticipation of a repeat experience. These can be applied in product development by asking research participants to consider repeating the same product experience. In doing so, the recent memory of the product experience leads naturally to feelings of desire or disgust about the anticipation of a repeat of the experience.

---

13. Stone and Sidel, Sensory Evaluation Practices

- Desire – Please rate how much you desire (if at all) to repeat this product experience (rate not at all to extreme desire).
- Disgust – Please rate how much you are disgusted (if at all) at the thought of repeating this product experience (rate not at all to extremely disgusted).

Text information from consumer stories and open-ended comments is often extremely rich in context, but difficult and time consuming to analyze. There are a number of strategies to convert qualitative text into quantitative verbatim summaries. The appraisal framework is a great framework for the conversion of qualitative text into quantitative metrics. The instant, specialty coffee experience from "Nora" is an example of how her story was converted into categorical measures (see Chapter 4).

These consumer measures form the basis for a more behavior-driven approach to product development. In building knowledge throughout the earlier phases of the product development process, the landscapes of consumer experience focus innovation teams, not only on the conceptual or tangible definition of the product, but also on the "whys" of consumer responses to experiences. This knowledge enables teams to understand what consumer metrics are most appropriate to be applied in generating consumer product insights, and to build models that lead to knowledge about the product, process, and quality landscapes of consumer experiences.

PRODUCT OPTIMIZATION ITERATIONS

As with other phases of the Product Development Cycle, the optimization stage of development is highly iterative. The rate and number of iterations are often constrained by a number of challenges associated with the producing of test food prototypes, including ensuring that they are safe for human consumption, and the organizational effort associated with fielding consumer studies to obtain feedback.

Product optimization follows the basic learning process used in other phases of product development. There is a phase of generating knowledge about the product landscape. This knowledge is typically used to refine the gold standard – considering refinements in adjusting to commercialization requirements and achieving reach in emotional impact across the target consumer range. These iterations cycle through activities of inspired creativity about how to overcome the challenges of meeting both the commercial and experience requirements, followed by activities of refining prototypes and making decisions. This leads to the stages of process optimization and process/quality control. The final specification is typically validated through intensive home-use trials.

There are two basic goals in product optimization. The first is to refine the gold standard by optimizing the combination of cues and food qualities to maximize emotional reach. This involves the development and use of sensory hurdles, which are achieved by holistic techniques that integrate sensory descriptive measures into consumer response measures. Within this landscape, optimum products are identified that achieve the greatest emotional impact, as defined by sensory hurdles. The second goal is to maintain a focus on delivering the experience as defined by this gold standard while incorporating the commercial requirements. This involves the development and use of consumer tolerances.

Product Landscapes

The strategy for product optimization involves building and using models that integrate consumer response to technical product development measures. These models become the basis for knowledge building about the product landscapes within which developers operate. In practice, sensory supports development by applying these models in three ways: (1) consumer response predictions and estimation; (2) model learning (e.g. importance drivers, optimization); and (3) generating specifications from consumer tolerances.

Landscape models are statistical (or heuristic) models that generate insights, to inspire and guide product development. The simplest consumer response models involve one product development measure (e.g. sweetness) and one consumer response measure (e.g. Liking). An example of this is shown in Figure 10.2, involving the relationship between sweetness in Gatorade and

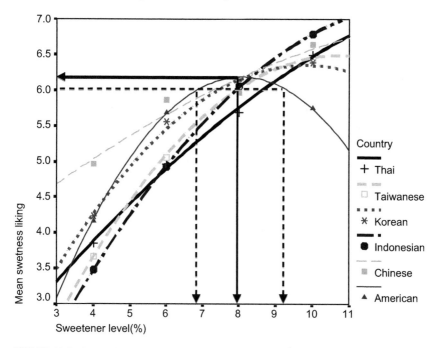

FIGURE 10.2 Mean Liking responses to varying levels of sweetness in Gatorade among consumers from different parts of the world and the lower and upper sweetener level from a 6.0 consumer tolerance level in Liking. *(Source: Chung, S. 1999. Cross-cultural sweetness preferences for a sports-drink. ms Thesis. Dept Food Science & Technology, Oregon State University, July 13, 1999. 183 pgs)*

consumer Liking from five different cultures.[14] In this case, the model fits with a strong curvilinear trend among American research participants. This model can be used to predict Liking for any level of sweetened Gatorade. Product development learns something from the model, in that a sweetness level 8.0 maximizes Liking with an average Liking rating of 6.2. Further, if the consumer tolerance level were set at 6.0, then this would translate to lower and upper "fiducial limits" (i.e. applying inverse regression) for sweetness with respective levels of 6.8 and 9.3. These could form the basis for a sweetness level specification for Gatorade sold to American consumers.

Product landscapes involve the building of models to predict consumer responses from technical product development measures. The basis of these models tends to be the sensory perceptual space that defines the consumer experience. This domain can be related to either technical product development measures (chemical, physical, and instrumental) to guide product developers, or to consumer measures based upon the appraisal framework to understand the impact of development decisions on behavior.

Holistic Preference Mapping

Preference mapping has been used by sensory researchers, since the 1990s, as a statistical technique to understand product landscapes. Early versions of preference mapping involved models that explained consumer Liking, based upon technical sensory descriptive measures. These methods were rooted in the internal and external preference mapping techniques developed by Chang and Carroll[15] and Carroll[16]. These techniques did not gain widespread use until after they were refined in the 1990s.[17,18] Later, partial least squares regression (PLS) was also applied to achieve preference mapping.[19]

Holistic research methods generate a richer data set than simple preference mapping can analyze. Consider the schematic in Figure 10.3. This shows a data structure of four different types of research information – information

14. S. Chung, Cross-cultural Sweetness Preferences for a Sports-drink, MS Thesis, Department of Food Science & Technology, Oregon State University, July 13, 1999, pp. 183

15. J.J. Chang, J.D. Carroll, How to Use MDPREF, a Computer Program for Multidimensional Analysis of Preference Data. Unpublished report, Bell Telephone Laboratories, 1968

16. J.D. Caroll, Individual Differences and Multidimensional Scaling, in R.N. Shephard, A.K. Romney and S.B. Nerlove (Eds.), Multidimensional Scaling: Theory and Applications in Behavioral Sciences, vol. 1. Seminar Press, New York, NY, 1972, pp. 105–155

17. K. Greenhoff, H.J.H. McFie, Preference Mapping in Practice, in H.J.H. McFie, D.M.H Thomson (Eds.), Measurement of Food Preferences. Chapman and Hall, London, 1994, pp. 136–165

18. J.A. McEwan, Preference Mapping for Product Optimization, in T. Naes, E. Risvik (Eds.), Multivariate Analysis of Data in Sensory Science. Elsevier, Amsterdam, 1996, pp. 71–101

19. H. Martens, M. Martens, Multivariate Analysis of Quality: An Introduction. John Wiley & Sons, Chichester, 2001

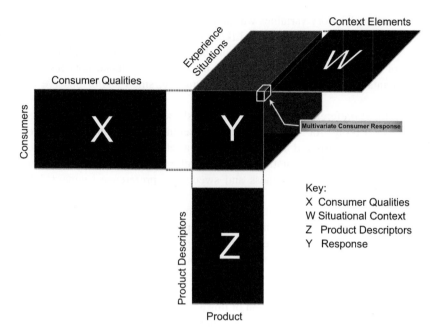

FIGURE 10.3 Structure of research data for the generation of preference maps based upon the appraisal framework.

characterizing the consumer, situational context, product qualities, and consumer response measures. These data sources comprise the basic building blocks for a causal model from the appraisal framework (Chapter 4). There is often not just one consumer response generated, but a multivariate set of responses from holistic research methods. These responses can be any of the consumer response measures discussed earlier, including difference judgments, choice or preference, self-reported measures of Liking products, self-reported measures of feelings about experiences or how a product makes one feel. They can also be qualitative responses.

These four data sources can be interconnected to form predictive models when research participants appraise the same products in similar contexts, using the same set of response measures. The traditional PLS preference map uses only the product descriptors (data set Z) to explain one consumer response measure, such as Liking. A technique called L-PLS was published by Plaehn and Lundahl in 2006[20] to enable a second data set (e.g. qualities about the consumers, as in data set X) to be integrated into an additional set of explanatory variables. In this paper, the value of integrating these

20. D. Plaehn, D.S. Lundahl, An L-PLS Preference Cluster Analysis on French Consumer Hedonics to Fresh Tomatoes, Journal of Food Quality and Preference 17 (2006) 243–256

additional explanatory variables was shown by contrasting traditional PLS to L-PLS, to explain the variation in Liking for different tomato varieties. The data set involved descriptive analysis on the firmness, aromatic qualities, and color of 16 tomato varieties. Consumer information included how research participants select tomatoes (i.e. do they squeeze, smell and/or observe the color).

A typical preference map, using PLS, is shown in Figure 10.4. It models the mean Liking for three "preference segments," groups of research participants that differed in their tomato variety Liking, as characterized through a classical segmentation analysis. Preference segments 1 and 2 were driven more by tomato juiciness and sweetness, preference segment 3 by texture.

In Figure 10.5, these insights take on a more behavioral perspective, by including behavioral information to help explain Liking differences among these segments. Segment 1 relies on the smell of tomatoes as a cue for sweetness and juiciness. Segment 2 relies on tomato shape, color and skin width as cues for the type of tomato most appropriate for different functional uses. Segment 3 relies on a tendency to squeeze tomatoes to test for firmness as a cue for freshness. The integration of this additional information gives deeper insights into why each preference segment "likes" one variety or another.

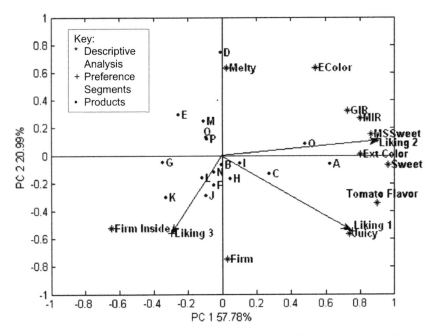

FIGURE 10.4 Simple two-way preference map of tomato varietal Liking among three preference clusters explained by sensory descriptive attributes.

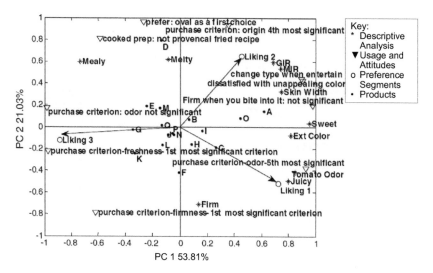

FIGURE 10.5 N-PLS preference map of tomato varietal Liking among three preference clusters explained by sensory descriptive attributes and usage and attitude data.

Segment 1 is driven by sensory pleasure, segment 2 by functional utility, and segment 3 by freshness.

Preference mapping can also be extended further to explain multivariate consumer responses – using a technique called Multiway N-PLS or mN-PLS regression.[21] Modeling to explain multivariate consumer responses helps not only generate insights into what behavioral drivers differentiate consumer segments, but also what emotions are formed as a result of product experiences. By knowing what emotions are at play, deeper insights into the drivers of consumer behavior can be inferred.

Consider again the case study for the development of a new, specialty, instant cappuccino coffee in which consumers evaluated five coffees – one of them being an organic cappuccino. Prior to consumer testing, an expert panel provided assessments about the packaging and packaged coffee drink qualities (potential cues) that characterized product differences. Consumer response information was generated from research participants who appraised each product in a random, sequential monadic order. They rated their feelings generated from each product experience on 12 emotions (Liking, Amusement, Satisfaction, Relief, Anger, Pride, Boredom, Surprise, Fascination, Stimulation, Desire, and Disgust) and purchase interest. Research participants with preferences for specific products, and strong emotional responses, were selected for one-on-one interviews.

---

21. D. Plaehn, D.S. Lundahl, Regression with Multiple Regressor Arrays, Journal of Chemometrics 21 (2007) 621–634

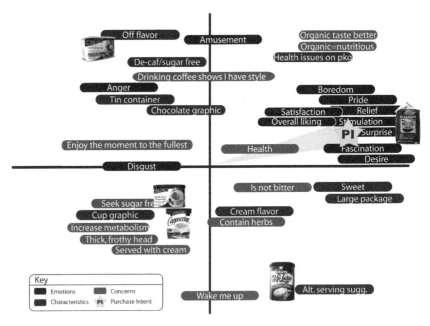

FIGURE 10.6 mN-Way preference map among four preference segments with regard to a multivariate set of responses (12 emotions attributes and purchase intent), explained by descriptive attributes and consumer concerns about their instant cappuccino experience. Showing only the predicted responses from the "organic-driven" behavior segment.

Research participant profiles, in purchase interest for these five coffees, were used to characterize consumers into four "behavioral segments." The differences between two segments ("organic driven" and "frothy driven") are shown, respectively, in Figures 10.6 and 10.7.

The mN-Way PLS technique[22] enables insights that go beyond what can be gleaned from a PLS or L-PLS preference map. Consumers exhibiting an organic behavior driver can be separated from those driven by the sensory qualities of an authentic cappuccino. The variation in multivariate response behavior among these five coffees was explained by both the consumer concerns and product characteristics (i.e. potential experience cues). As expected, those preferring the organic product were more concerned about health issues, but had different attitudes about organic products. They were less concerned with the authenticity of the cappuccino – and that the product be sugar-free. In fact, the presence of sugar-free on a product label was associated with self-reported disgust. The predominant feelings reported from the organic product were Surprise, Stimulation, Relief, Fascination, and Desire.

---

22. Plaehn, Lundahl, Regression with Multiple Regressor Arrays

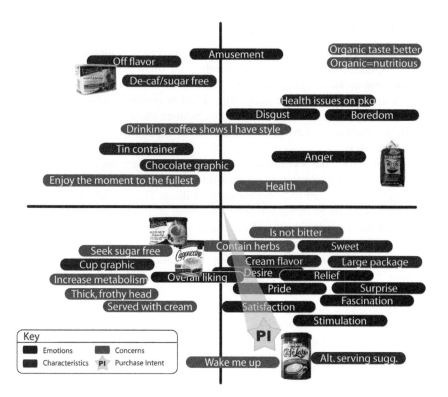

FIGURE 10.7 mN-Way preference map among four preference segments with regard to a multivariate set of responses (12 emotions attributes and purchase intent) explained by descriptive attributes and consumer concerns about their instant cappuccino experience. Showing only the predicted responses from the "authentic-driven" behavior segment.

These qualities were extremely different for those who preferred the more authentic cappuccino products. These research participants were characterized as being driven by functionality (Wake me up) and authenticity in product performance. The predominant emotions associated with their experience were Stimulation and Satisfaction.

Methods that are a hybrid of quantitative and qualitative research are proving invaluable in generating additional insights that cannot be attained by the most advanced statistical modeling techniques available. The qualitative one-on-one interviews add new information that can be integrated with the information generated from these preference maps. Characterizing "Nora" as being driven by authenticity (e.g. cued by frothiness), her story about an envisioned context of use (i.e. sharing with her girlfriends) deepens the understanding of the product landscape for product optimization. The cue of frothiness increases in importance as a differentiating product quality that can be measured and optimized through product development.

Optimizing the Gold Standard

In the design phase, a gold standard was developed to help identify the experience requirements. This was done mostly with small groups of consumers participating as co-designers with the research chef and package designers. This design translated the concept into tangible qualities, cuing emotional impact among the co-designers. In the development phase, this design is taken out to the market to generate broad consumer input, ensuring it will have emotional reach.

In Figure 10.2, the sweetness levels of Gatorade were varied to generate Liking responses among consumers from different countries. A model was fit to the American consumer response means for Liking. This model predicted that the optimum for the beverage was at 8.0 sweetness units. However, sweetness is but one of many contributing qualities and design elements that can contribute to Liking, or other measures of emotional impact.

Among target consumers there is often significant heterogeneity in response behavior to products with varying levels of sweetness and other qualities. Not only do consumers vary in what cues elicit emotional impact, but there is diversity in optimal levels of qualities such as sweetness that maximize emotional impact. In Figure 10.8, a schematic of a contour map is shown. Different shaded areas represent increasing levels of response behaviors to products varying in two sensory qualities (e.g. sweetness and sourness). Within the ranges of these two levels the maximum level of consumer response is shown in the lower left hand corner.

If a line of products were designed with respective gold standards in the design phase, then these products help capture broad differences in preference over the consumer target. However, there is often a need to optimize each product in the line, to ensure it is maximizing emotional impact for the greatest proportion of target consumers within its respective domain. Typically, there are many more than two qualities to vary. Contour plots of Liking, and other response measures over a more complex set of qualities, can be shown as two "dimensions" of sensory qualities – each dimension

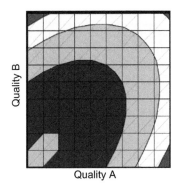

Quality B

Quality A

FIGURE 10.8 A simple two-dimensional contour plot of sensory qualities (horizontal and vertical axis) by consumer response (contours).

being different linear combinations of sensory qualities that best explains the response variation.

Consider the case study for the development of a dairy product. A company was looking to further develop a new prototype formula (prospective "gold standard") that had been designed in the laboratory into a winning product. This prototype, two "extreme" prototypes, and 11 competitive products were characterized through descriptive analysis. The qualities of all 14 prototypes and products were characterized using 41 descriptive attributes. These same samples were appraised by 235 target consumers in three geographically dispersed locations throughout the target market.

Preference mapping was applied to build a product landscape map covering the Liking responses of consumers to these diverse sensory experiences, using the method published by Plaehn.[23] This method is based upon generating individual response models for every research participant, and generating a distribution of predictions over a grid of "points" across the domain of different descriptive qualities.

In Figure 10.9, a contour map is displayed of mean Liking responses from these consumers over the first two dimensions of the descriptive differences among the prototypes and products. In the upper right hand corner of the contour plot, an optimum was found where Liking is highest. This optimum prediction is also associated with a set of sensory experience specifications. The gap between the sensory qualities of the prospective "gold standard" and the optimum sensory specifications provides a clear roadmap for developers for how to improve the gold standard on the basis of maximum Liking.

Contours, such as this, also can lead to focused development activities to achieve sensory hurdles. If the sensory hurdle is a "measured consumer response" of a 7.0 in Liking, then the optimization goal for development is to follow this roadmap to improve the mean Liking response from the current level of 6.5 to 7.0. However, if the sensory hurdle were defined as a 60% probability for a measured consumer response of 6.0 or higher in Liking, then a different contour would need to be built to guide development.

Consider the alternative contour map from the same data (Figure 10.10). In this map, the contours are the predicted probabilities that a randomly selected target consumer will rate various dairy products as 6.0 or higher. It is interesting to note that three regions in this map have equally high probabilities – the area around the competitive product C1, the "optimum" maximum mean Liking response, and right-most region of the contour map.

In optimizing the gold standard, it is important to consider the domain in which a product will play. In this dairy product case study, the roadmap

23. D. Plaehn, A Variation on External Preference Mapping, Journal of Food Quality and Preference 20 (2009) 427–439

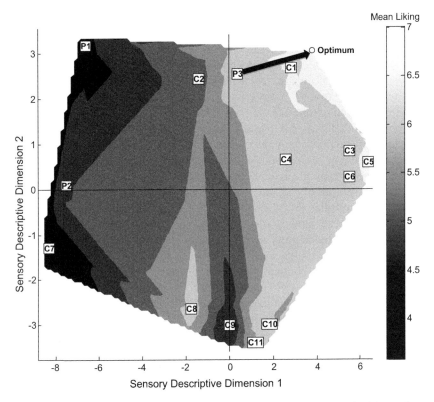

FIGURE 10.9 Preference map for mean Liking (contours) among 14 samples of a dairy product differentiated on 41 descriptive attributes (first two dimensions as horizontal and vertical axis) with optimum prediction within the prediction space.

to improve the prospective gold standard, by matching the predicted optimum, involves fairly subtle changes in formulation. The sensory differences in the far right or lower right corner of the contour map involve more extreme formulation changes. These alternative regions can be considered as different experience domains within which to optimize a gold standard.

Generating individual models for research participants, as a basis for preference maps, leads to new insights. Contours can be easily generated for different market segments of consumers or for different situational contexts, as shown in Figure 10.3. Further, incorporating regression effects other than sensory product qualities (i.e. consumer and context information) into multi-way PLS models enables insights to be generated, as to the importance of these effects in explaining preference.

Analyses such as this helps the innovation team gain more clarity in defining the experience target, refining the respective gold standard for a given

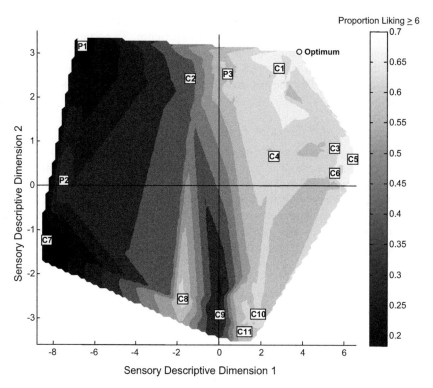

FIGURE 10.10 Preference map for predicted proportion of consumers scoring Liking at 6.0 or higher (contours) among 14 samples of a dairy product differentiated on 41 descriptive attributes (first two dimensions as horizontal and vertical axis) with optimum prediction within the prediction space.

sensory experience domain. It is equally important that the packaging and other aspects of the whole experience also become refined, to ensure that all components of the experience deliver on the promise of the concept and that these components are themselves in harmony. This can only be achieved through a more holistic approach to development.

Holistic Refinement

Optimizing the gold standard before developing the product leads to product optimization efficiency. It puts the focus on getting the experience right before tackling the challenge of incorporating commercial requirements that might lead to experience compromises, or tradeoffs. The first step in developing a commercial viable product is to consider how to meet this challenge, which is heightened in the food industry – owing to the complex interactions that often exist between package, packaged food, and messaging. These interactions make it difficult to make development decisions, without inadvertently impacting another aspect of the consumer experience. To account for these interactions when considering specific material, formulation or even messaging changes the innovation team needs to maintain a holistic perspective. This is best accomplished when the package and packaged food can be developed together.

Rapid Product and Package Testing

In Chapter 5, rapid iterative testing and adaptive research were introduced as emotions research methods. These are listed in the center of Figure 10.1 as part of the family of holistic research methods. These methods are holistic due to their application of holistic research designs, enabling insights into how changes in package components and food composition interact to impact the consumer experience. By incorporating these methods into the normal development iterations, the innovation team can quickly and efficiently refine products that deliver on the promise of the concept, with minimal tradeoffs or compromises.

Consider the case study of a microwavable, packaged side dish. The research protocol is summarized in Figure 10.11. The study involved 450

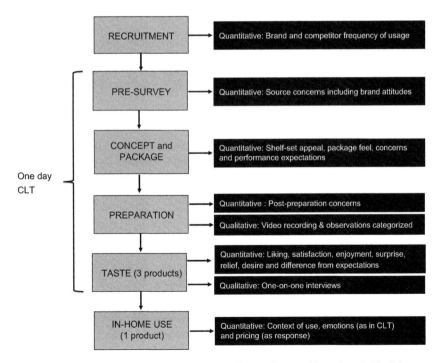

FIGURE 10.11 Holistic research design to refine a microwavable packaged side dish.

consumers, recruited as research participants from three markets. Five segments were recruited on the basis of category behaviors to participate in a central location study. They followed a protocol similar to how they might naturally use the product – a moments-of-truth holistic design. Participants filled out a pre-survey about their habits, source concerns, and brand attitudes. They then appraised packaging in a shelf set and off the shelf, answered questions and provided open-ended comments about their experience. They were asked to read the instructions, prepare the product and fill out questions about the preparation experience. This activity was also videotaped and behaviors were quantified in real time by a recorder. Next, three products were tasted in sequential monadic order with self-reported feelings rated.

The consumer response information was monitored in real time, at each central location, to select research participants for one-on-one interviews. Finally, participants were screened on the basis of having a high interest in purchasing the product, if it were available. As a result, 100 participants were invited to take one of the products home to try. These participants were surveyed, through a follow-up call, about their in-home experiences with the product.

This moments-of-truth design led to insights into the expectations generated from the package. Insights into how the packaging looked and felt when handled led to changes in package form. The package was not distinguished on the shelf, leading to changes in the package artwork. It was also observed that many consumers did not follow the instructions, even though they recalled that they had. These insights resulted in changing the instructions and package design, to avoid consumers burning their hands.

Satisfaction, enjoyment, surprise, liking and desire were the predominant emotions expressed as associated with the whole experience. Significant product differences were found on the basis of disappointment and enjoyment. Emotional impact was higher among heavy than light category non-users. Purchase intent was low for all products. The CLT results were confirmed through the in-home use testing, with additional insights gained about portion size.

This study demonstrates the value of a more holistic approach to product optimization. These research methods led to a wealth of information about how different aspects of the product could be improved. The ratio of insights gained, per cost of research, was extremely high compared to a series of smaller research studies to independently test refinements to package or packaged food. Subsequent development improvements resulted in the launch of a very successful product that has developed into the category leader.

Holistic research, using a hybrid of quantitative and qualitative research techniques, leads to insights into the complexities between different aspects of the consumer experience. Deeper insights were gained through a broader set of self-reported questions about emotional impact. While the product consumption experiences did not differ in Liking, there were significant differences found on the basis of enjoyment and disappointment.

Development Validation

As with the design phase, the validation step is intended to provide decision metrics about the performance of the product to achieve development goals. These goals are to develop a specification for a commercially viable, manufactured product, meeting the experience and commercial requirements.

Developing the Commercial Product

Thus far, in this chapter, only the first stage of development has been fully covered – product optimization. Validation can be achieved with a pilot plant product; however, this is typically done using a fully manufactured product. Therefore, prior to validation, the process of developing a commercial product must be completed.

When a new process is the critical path to aspects of the experience, this process must be validated to ensure that a manufactured product performs as expected. In these cases, the optimized and refined gold standards serve as a target for the process to match. Sensory research can apply difference testing or other metrics of consumer tolerance, to ensure that the target experience is not compromised.

The building of specifications is a lengthy process for most companies. Manufacturing Enterprise Requirements Planning (ERP) systems typically have product development modules to fully define the complete specification for a product. This includes specifications for processing and quality controls. These types of specifications often involve sensory research to apply consumer tolerances to models, interrelating consumer to technical development measures. As stated earlier, the building of these technical specifications goes beyond the scope of this book.

Validation Methods

Once a commercially viable product can be produced, validation serves to ensure it delivers the expected experience to consumers. This requires that the product be tested in the context of use with target consumers throughout the market. The behavior-driven approach to validation establishes success criteria that are not only consumer response-based, but also behavioral-based.

Behavior-driven validation criteria align with the behavioral goals of the innovation team. Behavior-driven validation can be either internal or external.

External validation tests the newly developed product against benchmark or competitive products. The criterion for success is behavioral.

Consider an example where the behavioral goal is to achieve a breakthrough product that disrupts incumbent use habits and establishes new habitual use for the new product. In this example, validation requires that three success criteria be met: (1) disruption; (2) trial; and (3) repeat. The disruption criterion provides proof that the developed marketing messaging engages the conscious mind of target consumers with new concerns that weaken their motives for habitual use of the incumbent. Traditional in-home use tests (iHUT) recruit for concept acceptors. Disruption testing is more elaborate in that it focuses on disrupting behavior, i.e. testing for a decrease in desire to continue habitual use of a competitive incumbent or seeking behavior to find an alternative.

The trial criterion provides proof that the package design and/or marketing messaging leads to selection behavior for the test product – over alternative choices. This behavior is best tested in mock shelf-sets or computer-aided mock-ups of choice sets. A key trial criterion for external validation is that a product be viewed as price-sensitive against the competition. This is often tested through Van Westendorp price sensitivity analyses (Chapter 8) on test and incumbent products to understand ranges of pricing (not too cheap and not too expensive). These ranges can then be used as a more holistic approach to test the impact of pricing on choice alternatives (trial versus the incumbent).

The repeat criterion provides proof that once the product has been tried there is a change in selecting or sharing behavior. Evidence for selecting behavior may be a greater desire to repeat the new experience than the old with the incumbent. Sharing behavior may involve evidence proving an increase in peer-to-peer chatter with positive sentiment. The Van Westendorp price sensitivity can also be included, after repeat use of the test product, to test the impact of pricing on repeat selection behavior, as in testing the trial criterion.

Internal validation typically makes sense only when there is no clear benchmark or competitive product. It involves monadic in-home use testing against sensory hurdles. The more emotional or behavioral the hurdle, the more predictive the hurdle is of the chance that a product will become a market success. As discussed in this chapter, once a product has been introduced into the marketplace, it must deliver a consistent experience in order for it to be successful.

The ultimate gauge of emotional impact is how a product leads to the formation of new habits. Proving this is in the domain of the final phase of product development – the tracking of product performance in the marketplace. This is covered in the final chapter of this book – The Innovation Company.

Key Points

- The goal for development is a specification that delivers the experience requirements constrained by commercial requirements.
- A commercial product must not only deliver the experience, but also meet various requirements such as conformance, reliability, durability, scalability, distribution, disposal and merchandizing.
- The development phase is characterized by its process stages of product optimization, process optimization and process/quality control.
- Process optimization and process/quality control are associated with consumer requirements; process optimization delivers differentiation.
- Sensory researchers are the champion for the consumer experience. Gold standards are a key tool for use by sensory in keeping product developers focused on delivering the experience.
- Sensory hurdles are metrics that guide development in meaningful ways and can be used as a basis to gauge success in achieving goals.
- Consumer tolerances are essential metrics to help define specification ranges; gold standards define the specification target.
- Sensory methods and techniques can be classified by what responses are measured (i.e. sensory affect or sensory perceptions), how they are measured (i.e. qualitative or quantitative) and in what context they are measured (i.e. laboratory, central location, in-home/in-context use).
- Consumer response measures in a behavior-driven development phase are those that can be associated with behaviors through the appraisal framework.
- In a holistic research context, Liking is not always a response to sensory pleasure. It is better defined as a non-descript, generalized measure of emotion projected on a product.
- Preference maps deepen insights into the product landscape within which products are optimized to achieve reach over the target market. Extensions of PLS (i.e. L-PLS and mN-PLS) enable behavioral insights to be generated by integrating different types of information.
- The integration of insights from preference maps with qualitative content further deepens behavioral insights into the product landscape.
- Refining the gold standard into an optimized product requires a holistic approach where package and packaged food are tested together within holistic research designs.
- Behavioral validation at the end of development involves generating evidence proving that developed product specification does in fact lead to behavioral change through repeat use.

Chapter 11

The Innovation Company

Innovation distinguishes between a leader and a follower.

Steve Jobs

INNOVATION LEADERSHIP

The difference between releasing a breakthrough, or a me-too product, into the marketplace depends on the desire of a company and its brands to lead. Steve Jobs – the CEO and co-founder of Apple – is one of the innovation giants of our time and his quote above equates innovation to leadership. Built into the heart of Apple is the desire to lead. They are an innovation company.

Founded in 1976, Apple has grown to be the largest technology firm in the world. In mid 2010, Apple overtook Microsoft to become the world's largest technology company with respect to market capitalization. With a $300 billion market value in January 2011, Apple is now the world's second largest company behind Exxon Mobil.[1]

Apple has grown due to its culture of innovation. Innovation is its lifeblood. It has found a formula for focusing on the continuous development of the next generation of products, the game changers of the marketplace. Their products break through the clutter of the competition because they have been designed to deliver emotional impact. In fact, Apple's products are so differentiated, they define new product categories.

The food industry is not dominated by an Apple. Instead, it is dominated by companies fighting to stay alive in a highly commoditized marketplace. The food industry has yet to find a formula to consistently cut through the clutter. Price sensitivity is high. Retailers are finding it easy to compete with name brand owners. The number of food SKUs in the marketplace has ballooned to over 30,000.

Yet, there are category leaders in the food industry, and these leaders have risen above their competition because they deliver product

---

1. http://en.wikipedia.org/wiki/History_of_Apple_Inc.

Breakthrough Food Product Innovation Through Emotions Research. DOI: 10.1016/B978-0-12-387712-3.00011-6

experiences. Bush Brothers dominates the canned beans cooking experience. Their brand and line of products is a household staple – a required side dish with the grilling experience. Kraft Macaroni and Cheese – the "blue box" – has owned a segment of the stove top cooking experience for over a generation.

Frito-Lay has led the snack food category for the past 40 years. Its success can be attributed to its focus on the consumer, and delivering to the market consumer experiences with emotional impact. As discussed in Chapter 2, Frito-Lay has stayed focused in the past 20 years on bringing to market "better for you foods." The company has further demonstrated its commitment to sustainability through the release of biodegradable packaging. They are leading by innovating.

P&G found a formula to transform itself into an innovation company, refocusing on markets and categories that have a long-term payoff. Yet, in spite of their successes, they were unable to weather the storm of the recession started in 2008. They were not nimble enough to develop new products quickly in challenging times.

P&G and other category leaders, such as Frito-Lay, have not been able to fully break out of the commoditization cycle – as companies such as Apple have been able to do. Upon leaving as the CEO in 2010, A.G. Lafley felt that his vision of transforming P&G into an innovation company was only 10% complete. What was still missing?

What is missing is an innovative, disruptive formula for consistent winning in the food industry. Apple has been able to become an innovation company because they found that formula and have been able to continually reinforce the value of that formula to its employees, shareholders, and customers. This chapter will complete the guide to achieving leadership, by applying the behavior-driven formula for success through innovation.

In the previous chapters, the discussion has been on the innovation team and how they can accelerate product development through focusing on different aspects of the target consumer experience. In this chapter, we will return to a discussion about innovation from the perspective of the market place. It will address the topics of how marketing and tracking market performance relates to innovation.

The world of digital marketing is opening up new vistas to track the emotional impact of product experience. By tapping into the river of information, companies are achieving new capabilities to not only sense the changing desires of the consumer, but also to learn about how products released into the marketplace, do in fact, generate emotional impact.

It is through the tracking and predicting of consumer behavior, resulting from new product launches, that companies can learn to improve their innovation and product development decision making. By increasing the base of knowledge available to innovation teams, landscapes can more quickly be understood. This base of knowledge also leads to new opportunities to improve upon the metrics used to make innovation decisions.

THE NEW MARKETING MIX

Traditionally, marketing has been the alpha and omega of the innovation process. Marketers contribute to the front end of innovation by playing a key role in the development of brand strategy – where to play and how to play the game to win. They are also responsible for the back end launch of the product into the marketplace – how to build awareness for a new product, promote it to the marketplace, and set competitive pricing. The conceptual design of a product becomes the basis for marketing communication through various channels during launch.

Traditionally, it has been up to the marketer to ensure that target consumers try a new product. The marketing strategy is to build awareness for a product's market position, such that there will be more opportunity for trial to occur. This strategy also assumes that if product development has done its job, then trial will naturally lead to repeat and market success.

This approach to marketing comes from the concept of the marketing mix. Introduced by Neil Borden in 1953, the concept of marketing mix has become a classic marketing textbook theme.[2] It was later defined, in 1960, by E. Jerome McCarthy as the "4 Ps" – Product, Place, Price, and Promotion.[3] The marketing mix fits well with a classic approach to getting to trial in a world ruled by rational thought. However, as has been a continuous theme throughout this book, the rational, conscious mind is only a fraction of the real story for what drives trial in the world today. The lion's share of the real story is the role of anticipation emotions – triggered by cues in the environment and from the product itself – and controlled by the unconscious mind of the consumer.

The real story leads to a new world of marketing that is much more integrated into the innovation process. The holistic perspective of behavior-driven innovation yields a new marketing mix of controls for the contemporary marketer to apply to achieve market success. It is much more consumer-centric. These new controls not only motivate selection behavior to impact trial, they motivate the "4 S's" of behavior.

A new marketing mix, based upon the appraisal framework, is shown in Figure 11.1. The goal of reaching desire is achieved through the development and implementation of an innovation strategy through a more behavior-driven innovation process.

The behavior-driven approach to innovation results in eight levers for use by marketers. These levers below all relate to the 4 Ps of the marketing mix, and integrate the marketer into all aspects of the innovation process.

---

2. http://en.wikipedia.org/wiki/Marketing_mix
3. http://en.wikipedia.org/wiki/E._Jerome_McCarthy

FIGURE 11.1 The eight behavior-driven levers that marketers can use to achieve the 4 Ps of the marketing mix.

Emotional branding leads to long-term, sustainable relationships by demonstrating the brand will work for the target consumer.

Strategic focus leads to innovation speed and accuracy by identifying consumer targets in the brand strategy phase and destinations in the discovery phase.

Digital messaging leads to seeking behavior by providing information channels easily found by consumers.

Disruptive promotion leads to disrupting old habits and introducing the new product conceptually by increasing negative anticipation emotions such as fear/disgust for the incumbent, while increasing positive anticipation emotions such as hope/intrigue through cues built into packaging and marketing communications.

Depletion reduction leads to the lowering of barriers to trial by easing the cost-, time-, and energy- "of-entry." When consumers are depleted of money, time, and energy they tend to behave habitually (emotionally), relying on their unconscious mind.

Environmental priming leads to habitual behaviors when sensations from the environment are associated with desirable, new product experiences (e.g. Taco Bell's bell sound).

Experience differentiation leads to desire when consumers use and/or consume products that deliver what they seek.

Social networking leads to sharing behavior by providing communication channels that consumers can easily use.

LEARNING FROM EXPERIENCE

In this book, the eight phases of innovation learning have been represented as two interconnected cycles – strategy development and product development (see Figure 11.2). We have so far introduced each of these eight phases quite linearly (Chapters 6–10) – starting with business strategy and ending in development phase. To consider these phases as interconnecting cycles, there needs to be a learning loop where knowledge is built within phases and used, in not only subsequent, but future innovation and development projects. Development of innovation strategy that has knowledge built in must somehow be used for development of future business strategy. In turn, knowledge from development must somehow be used for future discovery. This is the new frontier of behavior-driven innovation.

One way to envision knowledge transfer from one phase to the next is through the different consumer landscapes that must occur from phase to phase. In the early phases of strategy development, knowledge was built about brand landscapes – how consumers are changing in their attitudes about brands. In the later phases of strategy development, knowledge was

FIGURE 11.2 The interconnection in learning cycles between the development phase of the Product Development Cycle and the innovation strategy phase of the Strategy Development Cycle.

built about food product category landscapes – how consumers are changing their behaviors within food product categories. When a product is introduced into the marketplace, its performance can be applied to learn more about its impact on the brand and food product category behaviors landscape. Other types of landscapes were discussed throughout the Product Development Cycle. At the front end of product development, knowledge was built about opportunity landscapes – what behavioral white space exists with regard to contexts of use and concerns within product categories. In the conceptual design phase, knowledge was built about concept landscapes – what cues have been built into existing food product concepts within categories that lead to anticipation emotions. This was followed by the product design phase where knowledge was built about product design landscapes – what cues have been built into food product that elicit experience emotions. Finally, in the development phase knowledge was built about the product landscape – what technical product development measures interrelate to consumer responses in the form of sensory hurdles and consumer tolerances.

When a product is introduced into the marketplace, its performance can be applied to extend learning about the opportunity, concept, product design, and product landscapes. The building of landscape knowledge, by tracking the performance of launched products, is the new frontier of behavior-driven innovation. By building landscape knowledge from tracking information, the linear product development process is reshaped into two interconnected learning loops of strategy and product development. This reshaping of the innovation and development process leads to new thinking about how to track product performance, and to develop more predictive decision metrics throughout the innovation process. It also leads to opportunities to improve the success rates of future innovation and development projects.

Behavioral-Based Tracking

Various metrics have historically been used to track the performance of newly launched products. The most common set of metrics include point-of-sale purchase statistics and research into repeat behavior associated with test markets. This includes a wide range of research methods commonly known as shopper insights research. Learning tends to be based upon classic trial and repeat data graphed over time. Trending analyses sometimes forecast future performance based upon prediction models.

Tracking food product performance through the behavior-driven perspective starts with the appraisal framework. In Figure 10.3, consumer response data were visualized as a three-dimensional cube: consumers, products, and appraisal contexts. Throughout the product development process, this cube is filled with consumer response data as research participants appraise brands, contexts, themes, concepts, protocepts, prototypes, and products. This cube continues to be filled after a product is launched through tracking studies.

The future of behavioral-based tracking will be through the application of the appraisal framework to develop sensory hurdle models – predicting from consumer response data the probability for a consumer behavior. This was introduced as one of the four types of sensory hurdles in Chapter 10. Research is currently under way to develop these models to predict behaviors of importance to the success of a product from standard consumer response data. These models are envisioned to be applied to predict actual individual behaviors from consumer response information within landscape models.

Tracking data provides not only the basis to gauge how well an product introduced into a market is performing, but also to "train" behavioral models by interrelating consumer response data to actual consumer behaviors. Recall the dairy product case study. In Figure 10.10, landscape maps were generated for the proportion of consumers who responded with a 6 or higher in Liking. A similar response surface could have been generated using a function of several consumer response measures. Further, that function could have been built from a prediction model with inputs being several consumer response explanatory variables, and the output being an actual behavior.

The application of such a model has implications to extend learning for any of the landscapes, throughout the product development process. Different landscape models are envisioned to be built with different "domains." These domains would include different explanatory variables relating to the three dimensions of the consumer response cube: an "object" (e.g. brand, theme, concept, protocept, prototype, or product), the consumer (e.g. concerns and other qualities), and context (e.g. where, when, and with whom appraised). These are envisioned to form the basis for a new generation of behavioral-based decision metrics throughout the innovation and development process.

Decision Metrics

The second frontier of behavior-driven innovation is to apply models trained from tracking information to develop better decision metrics. The value of any predictive model is in its extensibility to accurately predict or forecast performance. The food industry has relied on a small number of research suppliers to provide decision metrics as forecasts of volume and market share. These models forecast market statistics, based upon averaged consumer responses such as purchase intention, liking, and uniqueness. Also factored into these models are marketing measures, such as advertising and promotion spending, and other classic marketing criteria associated with the 4 Ps of the marketing mix.

These models have been criticized as having limited domains of accurate predictability – especially for novel products. Jamieson and Bass[4] found that less than 50% of research participants who responded with a top 2 box (i.e. a 4 or 5 rating on a five point scale) in purchase intent rating actually do try the product. These results are further supported by research on consumer packaged goods products reported by Young et. al (1998[5]). Hoeffler (2003[6]) provides convincing evidence that standard consumer response measures are predictive of behaviors for "incrementally new products" (INPs), but fail to be predictive for "really new products" (RNPs). He posits and provides support for a theory that INPs predict well, since they are differentiated on the basis of functional and utilitarian product qualities. Novelty tends to differentiate in other quadrants of the product experience.

The landscape models discussed in Chapter 10 are based upon individual models. They have been individually validated, which provides a basis for extensibility and accuracy. In the dairy product case study, a predictive landscape could have been generated for any collection of consumers engaged as research participants. Landscape model predictions would hold for any dairy product – including competitive products and brands that were previously characterized or might be characterized in the future.

LEADING THROUGH A BEHAVIOR-DRIVEN FOCUS

An innovation company is not just consumer-centric, it is behavior-driven. It strives to disrupt markets by manufacturing brands that deliver breakthrough product experiences. This objective is met by focusing on achieving behavioral outcomes through innovation. It uses the emerging field of emotions research to accelerate innovation team learning for how to deliver to market disruptive products. This enhances the collective intelligence of the innovation team. It helps to focus teams on delivering product experiences that consumers seek from brands that they trust. It is applied throughout the Product Development Cycle from discovery to development. It is also applied to consider how to launch products and track their performance.

The behavior-driven innovation story includes a number of recurring themes. These themes help define what it means to be "behavior-driven" – i.e. what is an innovation company. There are eight basic principles of behavior-driven innovation: (1) brand manufacturing; (2) nimbleness; (3) engagement;

4. L.F. Jamieson, F.M. Bass, Adjusting Stated Intention Measures to Predict Trial Purchase of New Products: a Comparison of Models and Methods, Journal of Marketing Research 26 (3) (1989) 336–345

5. S. Hoeffler, Measuring Preferences for Really New Products, Journal of Marketing Research 40 (4) (2003) 406–420

6. M.R. Young, W.S. DeSarbo, V.G. Morwitz, The Stochastic Modeling of Purchase Intentions and Behavior, Management Science 44 (2) (1998) 188–202

(4) holism; (5) adaptive learning; (6) integration; (7) emotional design; and (8) knowledge management.

Hitting the Home Run

Behavior-driven innovation with its eight basic principles provides a roadmap for food companies to achieve success – to lead in categories where they choose to play. This roadmap requires an unbending focus on being behavior-driven, i.e. to consider the behavior outcomes required to achieve innovation success, and then to work backwards in determining how to achieve those outcomes through innovation. Companies that are food industry leaders have learned that behavioral change requires a focus on being disruptive. To become disruptive, an innovation team must go through a process of discovery. Yet, so must the consumer.

By being behavior-driven, the innovation focus shifts thinking from developing and selling products that consumers will want to buy, to delivering to the market product experiences that consumers seek, select, share and desire to repeat-use as a new habit. This requires that consumers desire to change a number of different behaviors.

The change that consumers must go through is analogous to getting consumers around the four bases of the baseball game (see Figure 11.3). The application of emotions research methods and techniques, within the discovery phase, provide for a roadmap for get consumers to first base. This is achieved by disrupting sensing, habitual behaviors of consumers such that they will seek alternatives. Behavior-driven discovery successfully achieves this by helping innovation teams discover the behavioral white space (i.e. experience destination) that defines the opportunity with the biggest potential to change behaviors among target consumers.

To get consumers to second base, behaviors must shift from seeking alternatives – to selecting a new product for trial. This is accomplished through emotions research, within the concept design phase, that helps the innovation team in building concepts that fit with the experience destination consumers seek and have cues built into concepts that deliver high anticipation emotional impact.

To get consumers to third base, behaviors must shift from selecting, to desire to repeat. This is accomplished by the design of products that deliver the promise of the concept. Through the design phase the promise is not only delivered, the high experience emotional impact is achieved by building cues into the packaging and packaged food. High experience impact leads to desire to repeat the experience.

To get consumers to home base, behaviors must shift from trials to new habitual use of a product. This requires the reinforcement of experiencing emotional impact and the establishment of associated environmental cues that signal and motivate habitual use of products. This is accomplished through the

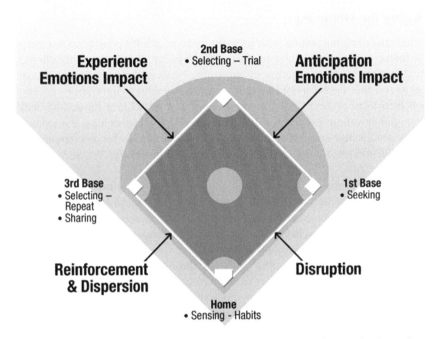

FIGURE 11.3   Behavioral changes involved in achieving breakthrough food product innovation.

development phase where the designed gold standard for the experience can be developed into a commercial product that consistently delivers the same experience.

Yet, getting one consumer to home base does not lead to success. By achieving high desire to repeat, the herding nature of consumers will naturally drive more sharing behavior. As more consumers become disrupted, the seeking out to be influenced by trusted peers will naturally drive more trial. The new marketing mix serves to accelerate these natural processes, by providing more ways to drive consumers around the bases.

Shifting the Innovation Paradigm

This shift in innovation thinking starts with a company's long-term focus on building relationships with consumers. By focusing on markets within which there are unfulfilled needs, innovation companies build relationships by manufacturing brands for the consumer. This is not a short-term, but a long-term focus. Brands must earn trust by demonstrating through their actions that they work for the consumer. Once trust is earned, consumers will more willingly hop on board to be taken to destinations that they seek.

This shift in thinking also requires giving significant control over to the consumer. This is what the prosumer is seeking. The metaphor of the brand as a vehicle is put into practice through more engaging sensory and marketing research techniques involving less interrogation, and more listening and dialoging. The innovation company is focused on knowing and anticipating the experiences that consumers seek. By tapping into the flowing river of information and applying emotions research methods, the landscape of the implicit mind of the consumer takes shape.

This shift in thinking embraces the data revolution with a framework for innovation team learning – the appraisal framework. This framework enables the diverse mix of quantitative and qualitative data to be integrated into insights. By staying focused on understanding the "whys" of the consumer landscape, insights become more relevant – getting to the "so what" of insights required by line managers, marketers, designers, and developers.

The result of the appraisal framework is a more efficient research process implemented across the cycles of strategy development and product development innovation. The appraisal framework serves to structure thinking about applying more holistic research to capture more complex and diverse data – providing confidence that testing bigger leads in the long run to lower costs, greater insights, and faster time to market.

In this way, emotions research fills a missing link in most company's research portfolio – a common focal point for all members of the innovation team. The framework represented by the Emotions Insights Wheel provides designers and developers with knowledge about the behavior drivers that can be built into food products that motivate consumers to take actions that drive success. It provides marketers with a new marketing mix of levers to disrupt markets. Lastly, it provides innovation managers with confidence in their decisions through a new generation of decision metrics.

Behavior-driven innovation is a roadmap to fulfill A.G. Lafley's vision for the innovation company. It provides the means for food companies to achieve leadership success- by focusing on changing consumer behavior, delivering product experiences that consumers seek, select, share, and desire to repeat. Looking to the future, the "perfect storm" within which the food industry finds itself is driving change at an unprecedented rate. The need to break through the chains of commoditization, and into the clear blue, has lead to the realization that innovation is the only viable, sustainable path forward. However, the approaches to innovation of the past will not suffice to lead companies into the clear blue. Those approaches are not nimble enough to help companies keep pace with the changing consumer. A more flexible paradigm must emerge to lead the way forward.

The basic theme of this book is that behavior-driven innovation provides the best path forward for food companies. The success stories told are but a few of the growing list adding proof to the value of applying this behavioral approach to innovation. Design in the food industry is becoming more commonplace with the incorporation of research chefs and packaging designers into the research process. Driven by a demand for more efficient and effective product development, the silos of the past are breaking down. The traditional linearity in the product development process is being replaced with a more holistic approach to product development.

Innovation managers and innovators are also demanding insights into the "whys" of behavior that have an answer to the question "so what." This is leading to heightened interest in emotions research to lead the path forward to breakthrough food product innovation. With its roots in the behavioral sciences, the appraisal framework is already a powerful framework to integrating the many streams of research information into faster, deeper emotions insights.

While new, emotions research is not a short-term trend. As behavioral and social science continues to develop deeper understanding into consumer behavior, this framework will continue to mature. This will result in new "best" practices, research methods, innovative techniques, and tools for the innovation team to increase research efficiency and effectiveness.

For these reasons, the food industry is still at the front end of this paradigm shift. This is evident by statements made by industry leaders, such as A.G. Lafley, that their respective companies still have a long way to go to become innovation companies. Many innovation teams are simply running at too fast a pace to put sufficient time and resources into rapid change. In spite of the heightened demand for emotions research, innovation teams are not able to find the time to try out and begin incorporating emotions research into their operations.

Yet, waiting for this framework to adopt is risky. In many ways, the future is now. Those not adopting this path forward will find the game changed *for* them - not *by* them. The forces of consumer change will continue to crowd their opportunity landscapes. The path forward is to embrace change, to break out into the clear blue through a behavior-driven approach – to innovation enabled by emotion research. This is the path that leads to breakthrough food product innovation.

Key Points

- Behavior-driven innovation provides a missing formula for food companies on how to be disruptive to markets.
- The appraisal framework leads to a new set of eight levers that marketers can use to impact market success. These are based upon the "4 Ps" of the marketing mix: emotional branding, strategic focus, digital messaging, disruptive promoting, depletion reduction, environmental priming, experience differentiation, and social networking.
- Tracking brand and product performance leads to the cycling nature of innovation strategy and product development. The frontier of behavior-driven

innovation is in the development of behavior-based tracking and decision metrics.

- Behavior-based tracking not only improves the gauging of success, but improves the predictive accuracy of landscape models used throughout the innovation and development process.
- Behavior-based decision metrics factor in different types of emotional impact than traditional metrics — leading to greater accuracy in forecasts, especially for novel products.
- The eight basic principles of behavior-driven innovation include brand manufacturing, nimbleness, holism, engagement, adaptive learning, information integration, emotional design, and knowledge management.
- The innovation company is not just consumer-centric, but behavior-driven. Focusing innovation teams on experiences that consumers seek leads to emotional impact driving the "4 S's" of behavior — Seeking, Selecting, Sharing and Sensing.

A Summary of the Basic Principles of Behavior-Driven Innovation

Basic principles of behavior-driven innovation	Description
Brand manufacturing	Innovation is the process by which brands are manufactured – i.e. born, protected, developed, extended, and evolved. Manufacturing of trustworthy brands leads to more opportunities to gain market share by building, extending, and protecting consumer relationships
Nimbleness	Innovation strategy requires a commitment to a long-term focus on winning in markets within which a company has decided to play, and a short-term strategy in how to adapt to playing the game. The result is greater nimbleness to respond to change without losing sight of long-term goals
Holism	Innovation companies tend to be holistic in two ways. First, the process for building and using knowledge throughout product innovation and development is more cyclic. Secondly, the research methods to learn are more iterative, using holistic research designs to gain insight into how the pieces and parts of products interact to form product experiences. This combination increases efficiency and effectiveness in bringing to market winning products
Engagement	Consumers are seeking more meaningful engagement with brands they trust will work for them through less one-way question and answer, and more interactive dialogue and active listening. Technology is enabling new platforms for these forms of engagement to drive innovation – further contributing to its efficiency and effectiveness
Adaptive learning	Holism, combined with engagement leads to research methodology, accelerates learning throughout the innovation and development process. These methods enable the adapting of holistic designs and dialogue protocols in response to iterative learning. This results in faster and deeper insights at lower costs

(Continued)

Basic principles of behavior-driven innovation	Description
Information integration	Insights are the result of integration of information using frameworks. Information integration through the lens of the appraisal framework results in behavioral insights. Behavioral insights provide the basis for the development of landscapes, forecasts, hurdles, and consumer tolerances that inspire creativity and guide decision making throughout innovation and product development
Emotional design	Design is the process of building into concepts, packaging, and packaged food cues that elicit emotional impact. Conceptual design is focused on eliciting anticipation emotions. Product design is focused on eliciting experience emotions. The greatest advances in near-term behavior-driven product innovation will be in the area of design

A Summary of the Goals, Guardrails, and Deliverables of Each Phase of Behavior-Driven Innovation

Behavior-driven innovation phases	Goals	Guardrails	Deliverable
Strategy development	Direct the innovation focus through a short-term innovation strategy that will build brand relationships with a target consumer	Long-term business (where and how to play), the brand and its strategy to evolve the brands (current and envisioned evolving brand equities, target markets, why to innovate)	The identification of the target consumer in behavioral terms (what behaviors to be changed) and the determination of what type of innovation (incremental, disruptive, platform) will achieve long-term business and brand goals
Discovery	Discover the opportunity for an experience that target consumers seek and branded products will deliver, thereby changing their relationship to the brand	Brand, brand equities, target market, target consumer and technical or market-driven platforms, innovation strategy	The identification of the experience theme that offers the greatest opportunity to change behaviors that will reach across the consumer target
Conceptual design	Define the product concept(s) for the product line that collectively achieves reach in emotional impact for the anticipation of the experience across the consumer target	The brand, brand equities, target market, target consumer, technical or market-driven platforms, innovation strategy, and experience theme	The identification of the concept(s) for each product in a product line that promises the experience that will reach across the consumer target
Product design	Design the product requirements for each product in	The brand, brand equities, target market, target	The identification of experience requirements for

(Continued)

Behavior-driven innovation phases	Goals	Guardrails	Deliverable
	a product line by translating the concepts into package and packaged goods to achieve reach in emotional impact for the holistic experience of the product across the consumer target	consumer, technical or market-driven platforms, innovation strategy, experience theme, and cues and expectations from the concept elements	each product that will create the experience that will achieve emotional impact reaching across the consumer target
Development	Develop the product specification for a commercially viable, manufactured product meeting the experience and commercial requirements for each product in the product line which collective achieve reach in desire for consumer trial and repeat of the experience across the consumer target	The brand, brand equities, target market, target consumer, technical or market-driven platforms, innovation strategy, experience theme, cues and expectations from the concept elements, and product experience and commercial requirements	The identification of specifications for each product that will deliver the experience that changes behavior required for success reaching across the consumer target
Tracking	Deliver the product to a test market and track the change in behavior associated with the product to forecast market impact (brand equity, market growth, share, and share take-away)	The brand, brand equities, target market, target consumer, technical or market-driven platforms, innovation strategy, experience theme, cues and expectations from the concept elements, and product experience and commercial requirements, test market domain, and launch strategy	The identification of market impact from launching the product(s) including forecasts in brand equity, market growth for the brand, brand share, and share take-away from competitors

Index

FIGURE 1.1 The gap between consumer and brand.

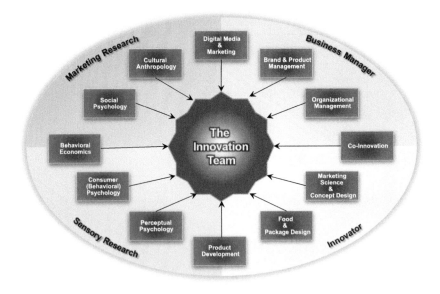

FIGURE 3.1 The multidisciplinary nature of innovation within the innovation team.

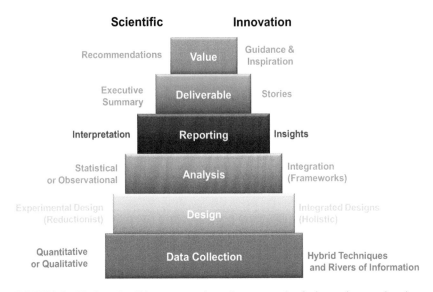

FIGURE 3.2 The Learning Cake – a comparison of two approaches for innovation team learning. On the left is a traditional (scientific) approach. On the right is a more contemporary integration approach.

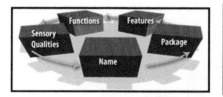
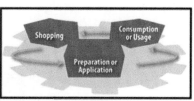
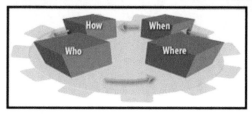

FIGURE 3.3 The three types of holistic research design: holistic product design (upper left), holistic moments-of-truth design (upper right), and holistic usage case design (lower).

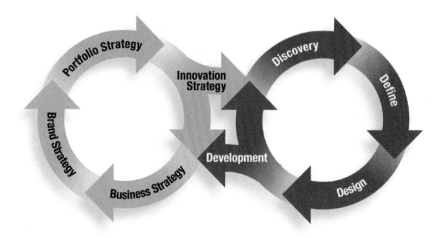

FIGURES 3.5 The Behavior-Driven Innovation Process with two interrelated learning cycles: the Innovation Strategy Development Cycle (left side) and Product Development Cycle (right side).

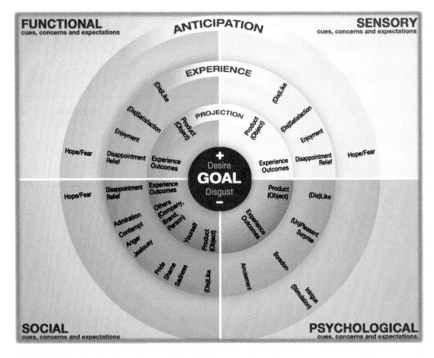

FIGURE 4.2 The Emotions Insight Wheel™ characterizing the expanded appraisal framework of anticipation and experience emotions associated with the four dimensions of product experience.

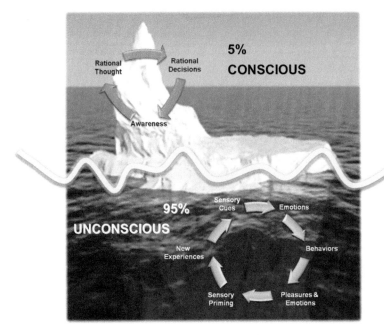

FIGURE 5.2 Interactions between the unconscious and conscious mind with regard to consumer behavior.

FIGURE 5.3 The Flavorcreator application displayed on Facebook used in "crowdsourcing" methodology for the Vitaminwater brand.

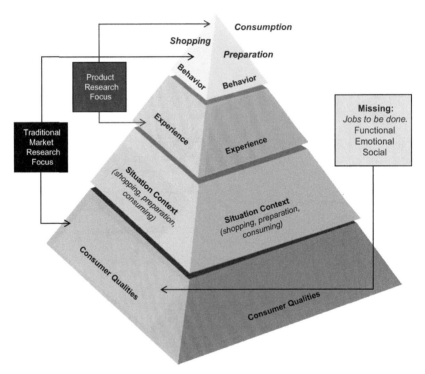

FIGURE 5.4 The Behavior Pyramid™ with the four levels of learning for consumer behavior (consumers, situational context, experience, and behavior).

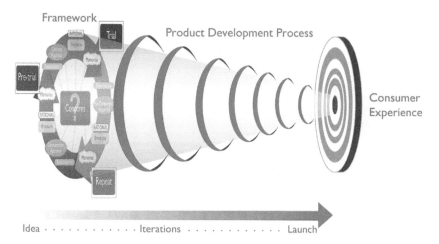

FIGURE 5.5 Holistic product development, as a highly iterative process of learning, to hit the target with a consumer experience that achieves emotional impact.

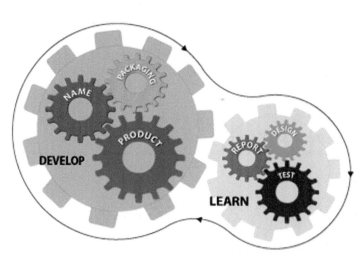

FIGURE 5.6 The interconnectedness between holistic product development (left side) and holistic research (right side).

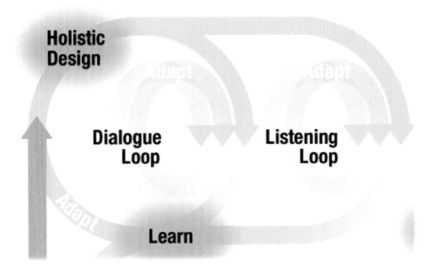

FIGURE 5.7 Adaptive research design where iterations are used to adapt the protocols for listening and dialoging, and/or to adapt stimulus within the holistic research design.

FIGURE 6.1 The Innovation Strategy Development Cycle with its four phases of Business Strategy, Brand Strategy, Portfolio Strategy, and Innovation Strategy.

FIGURE 6.2 The "before" and "after" logos for ConAgra Foods. *(Source: http://logos.wikia. com/wiki/ConAgra_Foods)*

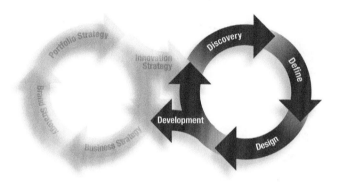

FIGURE 7.1 The Product Development Cycle with its four phases: Discovery, Define, Design, and Development.

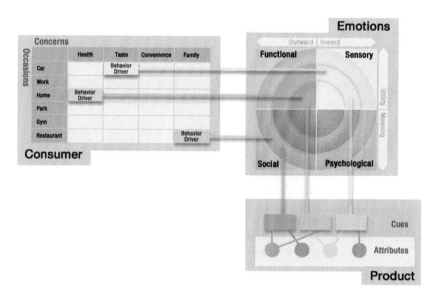

FIGURE 7.2 A knowledge map in the form of a brandTrace used for understanding the discovery landscape, with behavior drivers and associated cues mapped into the Emotions Insight Wheel.

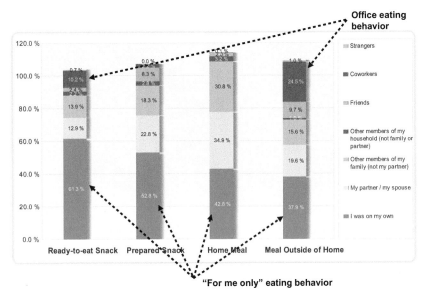

FIGURE 7.3 Results of activity-based research into food consumption behaviors for four types of foods (ready-to-eat snacks, prepared foods, home meals, and meals eaten outside the home).

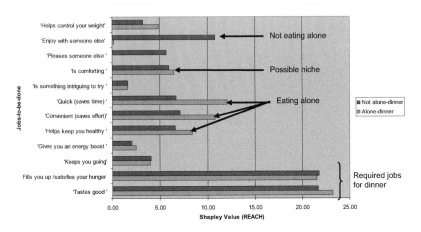

FIGURE 7.4 Differences in importance for various jobs-to-be-done by "at home" meals when eating alone or with someone else.

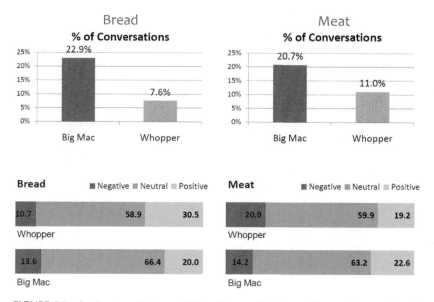

FIGURE 7.5 Sentiment analysis provided by Conversation from online social networks with regard to the bread and meat from Whopper (Burger King) and Big Mac (McDonald's Corporation).

How to Read a Theme Score Card

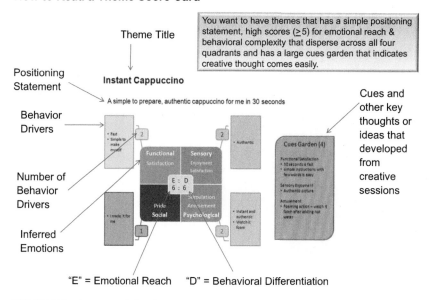

FIGURE 7.6 A scorecard for themes generated to characterize potential consumer product experiences.

Instant Cappuccino

A simple to prepare, authentic cappuccino for me in 30 seconds

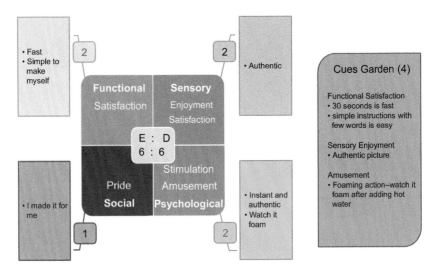

FIGURE 7.7 The "Instant Cappuccino" theme developed for the "Nora" archetype.

Secret Barista

My easy, fast, secret way to make an authentic cappuccino for friends

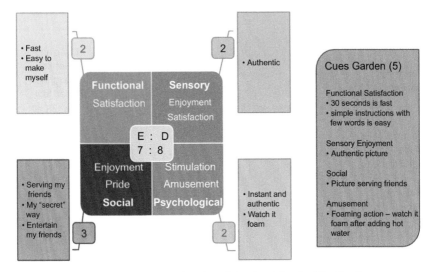

FIGURE 7.8 The "Secret Barista theme" developed for the "Nora" archetype.

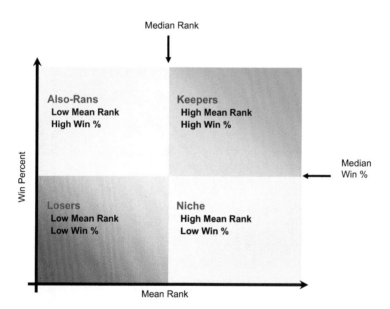

FIGURE 8.3 Quadrants of a choice-based analysis with ranking of winners.

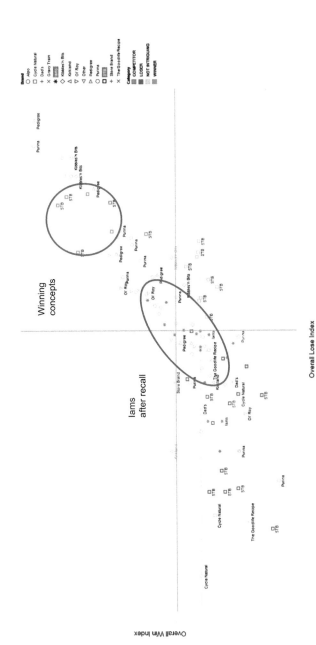

FIGURE 8.4 Results of choice-based analysis using maximum difference scales comparing new dog food concepts (STB) against competitive products. Six test concepts scored well against the top brands. Iams scored unusually low after a product recall.

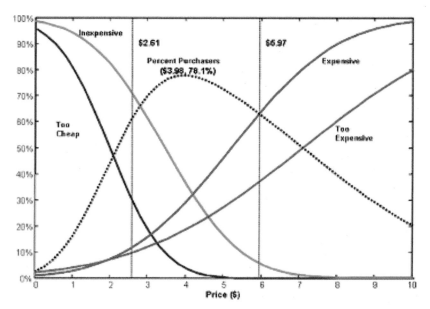

FIGURE 8.6 Van Westendorp approach to concept validation by using pricing as a measure of price-value.

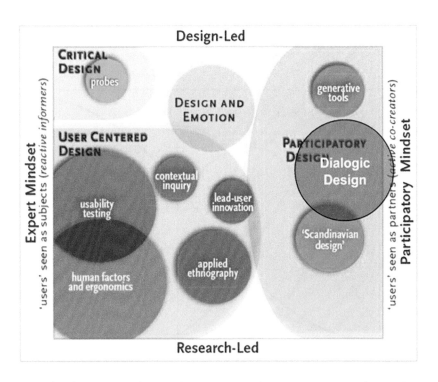

Topography of Design Research. Liz Sanders, Design Research in 2006. *Design Research Quarterly.*

FIGURE 9.3 Categorization of approaches to design based upon who participates (consumers or experts) and who leads the design effort (designers or researchers). Within the more participatory design methods there is an emerging area termed "dialogic design". *(Source: Peter Jones, "Transforming Contexts in Design Research," http://www.melodiesinmarketing.com/2009/07/25/ transforming-contexts-design-research-thinking-peter-jones-presentation/)*

Printed and bound by CPI Group (UK) Ltd, Croydon, CR0 4YY

08/05/2025

01864840-0001